Essentials of
Plant Physiology

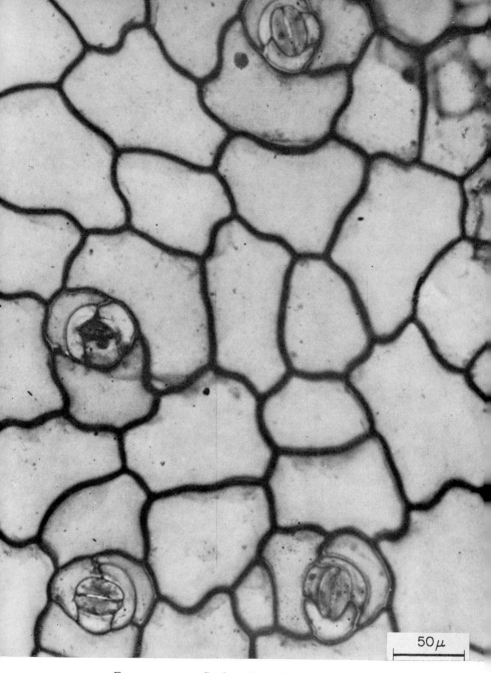

50μ

FRONTISPIECE. Surface view of stomata of Privet

Essentials of Plant Physiology

Second Edition

by

G. A. Strafford

Temple Moor School, Leeds

HEINEMANN EDUCATIONAL BOOKS
LONDON

Heinemann Educational Books Ltd

LONDON EDINBURGH MELBOURNE AUCKLAND TORONTO
HONG KONG SINGAPORE KUALA LUMPUR
NAIROBI JOHANNESBURG IBADAN
NEW DELHI

ISBN 0 435 60851 7

© G. A. Strafford 1965, 1973
First published 1965
Reprinted 1967
Reprinted with corrections 1968
Reprinted 1970
Second Edition 1973

Published by Heinemann Educational Books Ltd
48 Charles Street, London W1X 8AH
Printed in Great Britain by
Butler & Tanner Ltd, Frome and London

Introduction

For many years there has been an obvious need for a plant physiology text at a fairly elementary level in which the basic facts of the subject are given as concisely as possible. It is hoped that this book will remedy this deficiency. The necessary chemistry and physics have been kept to a minimum, but even so these subjects must be studied to at least Advanced Level G.C.E. if the student is to gain any real understanding of plant physiology—the days when Biology was a suitable subject for the weaker science student have long since passed! The subject-matter contained in this book should be adequate for Advanced Level in both Botany and Biology and candidates for special papers may also find it useful.

G. A. S.

Introduction to the Second Edition

The opportunity has been taken to include new material on photosynthesis, plant hormones and photorespiration and to bring up to date enzyme nomenclature.

G. A. S.

Skipton
March 1973

Acknowledgements

Quite apart from the assistance received indirectly from the authors of other text-books, specialist reviews, and original papers, I would like to make particular mention of the help given by a number of people.

In the first place Dr D. H. Jennings of Leeds University read the entire original manuscript and made many useful suggestions. Secondly the members of my third-year sixth form at King William's College, Isle of Man, in 1963–4—J. K. Brownsdon, A. R. Cannell, A. D. Garner, A. J. Rees, and W. R. Tingey—willingly allowed me to use them as guinea-pigs and, as a result, several modifications were made to the text.

Finally thanks are due to Mr Brian Bracegirdle for the production of the photomicrographs, Mr A. W. Iles for his skill in producing the final line drawings from my roughs and Mr Hamish MacGibbon for his endless patience in seeing the book through the various stages in its production.

G. A. S.

Isle of Man
August 1965

Contents

		PAGE
	List of plates	viii
1	Biochemical Principles	1
2	Diffusion and Osmosis	31
3	Transpiration and the Transpiration Stream	54
4	Translocation	70
5	Photosynthesis	83
6	Nitrogen and Fat Metabolism	102
7	Mineral Nutrition	123
8	Respiration	133
9	Plant Hormones I	153
10	Plant Hormones II	184
11	Germination and Growth	201
	Index	221

List of Plates

Surface view of stomata of privet *Frontispiece*

PLATE 1*a*. Section of *Syringa* leaf *facing page* 24

    1*b*. Section of *Ranunculus* root, to illustrate wide cortex of a root

PLATE 2*a*. Section of *Ranunculus* root, to show the endodermis and neighbouring cells   25

    2*b*. Section of *Iris* root

PLATE 3*a*. Section of *Curbita* stem   56

    3*b*. Section of *Helianthus* stem

PLATE 4*a*. Section of *Pinus* root   57

    4*b*. Section of *Syringa* leaf

1

Biochemical Principles

Carbohydrates

Carbohydrates may be defined simply as compounds which consist only of carbon, hydrogen and oxygen, with the latter two elements present in the same ratio as in water. Although this definition is adequate for an elementary treatment of the subject it is customary to consider closely related compounds as carbohydrates even if there are other elements present, or if the ratio of hydrogen to oxygen is not equal to two, as in desoxyribose (see page 3).

The simplest carbohydrates are the *Monosaccharides*, which can be represented by the general formula $C_nH_{2n}O_n$. These are further classified according to the number of carbon atoms, e.g. if:

$n = 2$—dioses $(C_2H_4O_2)$, e.g. glycollic aldehyde
$n = 3$—trioses $(C_3H_6O_3)$, e.g. glyceraldehyde, dihydroxyacetone
$n = 4$—tetroses $(C_4H_8O_4)$, e.g. erythrose
$n = 5$—pentoses $(C_5H_{10}O_5)$, e.g. ribulose, xylose
$n = 6$—hexoses $(C_6H_{12}O_6)$, e.g. glucose, galactose, mannose, fructose
$n = 7$—heptoses $(C_7H_{14}O_7)$, e.g. sedoheptulose

An alternative method of classification depends on the presence of either an aldehyde or a ketone group in the molecule (see below). The former are known as aldoses and the latter as ketoses.

The triose glyceraldehyde, particularly in the form of its phosphate ester, plays an important part in plant metabolism. The central carbon atom is asymmetrical and as a consequence shows optical isomerism, i.e. depending on the arrangement of the group attached to the central carbon atom, two isomers are possible which can be distinguished by their ability to turn polarized light either to the left or to the right. When the light is rotated to the left the compound is said to be laevorotatory, and its name is prefixed by the letter *l*: when, on the other hand, the polarized light is rotated to the right, it is known as dextrorotatory and is prefixed by the letter *d*. In the case of glyceraldehyde:

```
      CHO                          CHO
       |                            |
     HOCH                         HCOH
       |                            |
     CH₂OH                        CH₂OH
  l Glyceraldehyde            d Glyceraldehyde
```

These two compounds are the parent molecules of the higher monosaccharides. Since it does not follow that the higher monosaccharides rotate polarized light in the same sense as the parent triose, the capital letters L or D are used to denote which was the parent triose (*l* or *d*) and the signs (−) and (+) whether the compound is laevo- or dextrorotatory.

As an example, the case of glucose can be considered. This can be related either to *l* phosphoglyceraldehyde, in which case it is known as L glucose, or to *d* phosphoglyceraldehyde, i.e. D glucose. Only D glucose is found in plants and it is dextrorotatory, viz. D(+) glucose.

The formulae of the naturally occurring hexoses are given below. Glucose, mannose and galactose each contain an aldehyde group (i.e. are aldoses) and fructose contains a ketone group (i.e. is a ketose).

```
    CHO            CHO            CHO          CH₂OH
     |              |              |             |
   HOCH           HCOH           HCOH           CO
     |              |              |             |
   HOCH           HOCH           HOCH          HOCH
     |              |              |             |
   HCOH           HOCH           HCOH          HCOH
     |              |              |             |
   HCOH           HCOH           HCOH          HCOH
     |              |              |             |
   CH₂OH          CH₂OH          CH₂OH         CH₂OH
  D Mannose     D Galactose    D(+) Glucose   D(−) Fructose
```

The presence of either an aldehyde or a ketone group results in the well-known reducing properties of the monosaccharides, e.g. the reduction of cupric to cuprous oxide in Fehling's test, i.e.

$$CuSO_4 + 2NaOH \rightarrow Na_2SO_4 + Cu(OH)_2$$
$$2Cu(OH)_2 + R \cdot CHO \rightarrow Cu_2O + R \cdot COOH + 2H_2O$$

Glucose and fructose are the two most important hexoses. Of the other monosaccharides,

Erythrose ($CHO \cdot (CHOH)_2 \cdot CH_2OH$)
Sedoheptulose ($CH_2OH \cdot CO(CHOH)_4 \cdot CH_2OH$)
Ribose ($CHO \cdot (CHOH)_3 \cdot CH_2OH$)
Xylulose ($CH_2OH \cdot CO \cdot (CHOH)_2 \cdot CH_2OH$)

are all involved in the resynthesis of the carbon acceptor in photosynthesis.

Ribose is also an important constituent of ribose nucleic acid and of coenzymes I and II.

Desoxyribose is an important derivative, formed by the loss of an oxygen atom, found in desoxyribose nucleic acid.

$$CHO \cdot (CHOH)_3 \cdot CH_2OH \cdot - \tfrac{1}{2}O_2 \rightarrow CHO \cdot CHOH \cdot CH_2 \cdot CHOH \cdot CH_2OH$$
$$\text{(Ribose)} \qquad\qquad \text{(Desoxyribose, } C_5H_{10}O_4 \text{)}$$

Note that in desoxyribose the ratio of H : O is not two.

Oligosaccharides are formed by the condensation of two, three or four molecules of monosaccharides. Considering hexose condensations:

$$2C_6H_{12}O_6 \rightarrow C_{12}H_{22}O_{11} + H_2O, \text{ i.e. a disaccharide}$$
$$3C_6H_{12}O_6 \rightarrow C_{18}H_{32}O_{16} + 2H_2O, \text{ i.e. a trisaccharide}$$
$$4C_6H_{12}O_6 \rightarrow C_{24}H_{42}O_{21} + 3H_2O, \text{ i.e. a tetrasaccharide}$$

Examples of a trisaccharide and tetrasaccharide are:

Raffinose, which is particularly abundant in beet and is formed by the condensation of galactose, glucose and fructose.

Stachyose, which occurs in the tubers of *Stachys tubifera* and is formed by the condensation of glucose, fructose and two molecules of galactose.

The disaccharides are of more general occurrence in plant tissues. They may be formed by the condensation of either two identical molecules of the same monosaccharide or of two different monosaccharides. Examples of such condensation are:

$$2 \text{ Glucose} \rightarrow \text{Maltose} + H_2O$$
$$\text{Glucose} + \text{Fructose} \rightarrow \text{Sucrose} + H_2O$$

The resulting disaccharide may either be a non-reducing sugar or it may have half the reducing power of the two constituent monosaccharides. The former case occurs in the condensation of glucose and fructose since the condensation is between the second carbon atom of the fructose and the first carbon atom of the glucose (i.e. between ketone and aldehyde groups). In the formation of maltose, on the other hand, the first carbon atom of one molecule is linked with the fourth carbon of the other molecule so that one aldehyde group remains free.

Polysaccharides are formed by the condensation of a larger number of monosaccharides—at least ten, although the number is usually very much higher than this. The polysaccharides are classified according to the type of condensed sugar—e.g. hexosans by the condensation of hexose molecules, pentosans by the condensation of pentoses and mixed polysaccharides resulting from the condensation of both hexoses and pentoses.

This classification can be carried further on the basis of the type of sugar involved. Thus hexosans are divided into:

(a) Glucosans, formed by the condensation of glucose molecules— e.g. starch, cellulose and glycogen.
(b) Fructosans, formed by the condensation of fructose: the principal example is inulin.

Only starch and cellulose will be considered in detail here. Other polysaccharides, either pentosans or mixed, form mucilages, gums and hemicelluloses.

Starch consists of two components, amylose and amylopectin. They may be present in varying amounts, although usually amylopectin provides at least two-thirds of the total.

Amylose consists of long chains of glucose units joined together as in maltose. The chain is unbranched and the number of units in a single chain vary from 200 to 1,000.

Amylopectin is also made up of chains of glucose, joined similarly but in addition branches occur by the formation of 1:6 linkages.

There are about ten glucose units between branches and from 15 to 20 forming free ends.

The structure of the amylopectin molecule can be represented diagrammatically as

$\times$ = αD glucopyranose
— = 1:4 linkage (as in maltose)
$\oplus$ = 1:6 linkage

It will be seen that by a continuation of such a process, an essentially laminated molecule will be formed.

An important diagnostic test for starch is the formation of a blue colour with iodine. This is due to the reaction of iodine with amylose forming a bright blue colour and with amylopectin forming a deep purple.

The enzymatic synthesis of starch was formerly considered to be under the control of a 'P enzyme' and a 'Q enzyme'.

By means of a series of phosphorylations, the P enzyme synthesized amylose from glucose molecules while the Q enzyme was able to form amylopectin from chains of about 20 glucose molecules. In addition it could catalyse the production of such chains by the hydrolysis of amylose. Most of the steps involved have been demonstrated *in vitro* but there is considerable doubt as to their significance *in vivo*—for instance, there is no evidence of any phosphorylase activity in the chloroplast where most of the starch synthesis occurs.

*The role of uridine diphosphate** (*UDP*) in the synthesis of sucrose was first demonstrated by Le Loir in 1953 and the *in vivo* synthesis of sucrose by a similar reaction sequence was shown by Edelman in 1959. The necessary enzymes were found to be present in the scutellum of wheat and he showed that when an isolated scutellum was allowed to absorb either glucose or fructose, then sucrose was synthesized by a combination with uridine diphosphate. As in the case of many other reactions described in this book, the reaction is essentially cyclical (fig. 1).

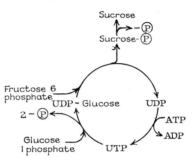

FIG. 1. Formation of sucrose

UDP reacts with ATP to form uridine triphosphate (UTP) and ADP. The UTP can then combine with glucose 1 phosphate to give uridine diphosphate-glucose (UDP-glucose) and this in its turn can react with fructose 6 phosphate to produce sucrose-phosphate and UDP. Apart from the regeneration of the UDP it is important to note that energy for the reaction is supplied by ATP.

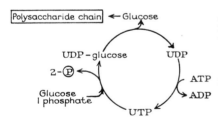

FIG. 2. Synthesis of amylose

By the addition of 'labelled' glucose (i.e. glucose containing radio-active carbon, ^{14}C) combined with UDP (i.e. UDP-glucose) to samples of starch prepared from beans, Le Loir was able to produce samples of starch in which some of the glucose units were radio-active. The starch samples used were contaminated with protein and this protein was found to be the appropriate enzymes. The cyclical reaction is shown in fig. 2, but the possibility of the P and Q enzymes being involved is not entirely ruled out since the

* UDP is a complicated molecule formed by the combination of uracil (a purine base) with two phosphate radicals and the pentose sugar ribose —i.e. a rather similar structure to ADP.

UDP-glucose system has only been demonstrated as a means of *extending* the length of the amylose chain: the P enzyme may be involved in the *ab initio* synthesis of amylose.

Cellulose is an important constituent of plant cell walls. It also consists of long unbranched chains but, unlike amylose, they consist of βD glucopyranose chains, i.e. the other optical isomer of glucose. They have a similar type linkage as joins the glucose units in maltose and each chain has from 1,000 to 3,000 units. The separate chains are held together in bundles by hydrogen bonds and short-range van der Waals forces.

Much less is known about the synthesis of cellulose. The role of UTP has been established for cellulose formation in at least one bacterium (*Acetobacter xylinium*) and in 1964 Elbein *et al.* reported evidence for the participation of guanosine diphosphate-glucose in cellulose synthesis by a cell-free suspension of Mung bean. Presumably GDP-glucose functions in a similar way to UDP-glucose.

Fats

Fats are esters formed by the condensation of three molecules of fatty acids with one molecule of the trihydric alcohol glycerol. If a fatty acid is represented by its general formula R·COOH, then fat formation can be written as

$$\begin{array}{ccc}
CH_2O\overline{H \quad HO}OCR & & CH_2OOCR \\
| & & | \\
CHO\ \overline{H + HO}OCR & \rightarrow & CHOOCR + 3H_2O \\
| & & | \\
CH_2O\overline{H \quad HO}OCR & & CH_2OOCR
\end{array}$$

It is not necessary that the three molecules of the fatty acid should all be identical—up to three different acids may be involved in condensation with one molecule of glycerol. The principal fatty acids found in plants are:

Butyric acid	$CH_3(CH_2)_2COOH$
Caproic acid	$CH_3(CH_2)_4COOH$
Palmitic acid	$CH_3(CH_2)_{14}COOH$
Stearic acid	$CH_3(CH_2)_{16}COOH$
Oleic acid	$CH_3(CH_2)_7CH:CH(CH_2)_7COOH$

Plant fats are usually liquid at normal temperatures (10–15° C) and are therefore referred to as oils. Whereas some fats enter into intimate protoplasmic structures, the majority are used as a form of food storage.

Interconversion of fats and hexoses is possible via glycerol (it will be recalled that the parent monosaccharide is glyceraldehyde, i.e the aldehyde of glycerol).

Proteins

Proteins are formed by the condensation of large numbers of amino acids, and this will be considered first.

The general formula of an amino acid is

$$R \cdot CH(NH_2)COOH \quad \text{or} \quad R—\overset{\displaystyle H}{\underset{\displaystyle \underset{\overset{\displaystyle N}{\overset{/\backslash}{H \quad H}}}{|}}{C}}—C\overset{\displaystyle O}{\underset{\displaystyle OH}{}}$$

Most of the naturally occurring amino acids are α amino acids, i.e. they have their amino (NH_2) group attached to the α carbon atom (the carbon atom *next* to the COOH group), e.g.

$$H—\overset{\displaystyle H}{\underset{\displaystyle H}{C^\beta}}—\overset{\displaystyle H}{\underset{\displaystyle NH_2}{C^\alpha}}—COOH$$

α Amino propionic acid
(Alanine)

$$H—\overset{\displaystyle H}{\underset{\displaystyle NH_2}{C^\beta}}—\overset{\displaystyle H}{\underset{\displaystyle H}{C^\alpha}}—COOH$$

β Amino propionic acid
(β Alanine)

Some of the principal α amino acids are:

Glycine	$CH_2(NH_2)COOH$
Alanine	$CH_3 \cdot CH(NH_2)COOH$
Valine	$CH(CH_3)_2CH(NH_2) \cdot COOH$
Leucine	$CH(CH_3)_2CH_2 \cdot CH(NH_2) \cdot COOH$
Aspartic acid	$COOH \cdot CH_2CH(NH_2) \cdot COOH$
Glutamic acid	$COOH \cdot (CH_2)_2CH(NH_2)COOH$
Ornithine	$CH_2(NH_2) \cdot (CH_2)_2CH(NH_2) \cdot COOH$
Citrulline	$H_2NCONH(CH_2)_3CH(NH_2) \cdot COOH$
Arginine	$HN{=}C(NH_2)—NH \cdot CH_2 \cdot (CH_2)_2CH(NH_2)COOH$

Tryptophan

$$\text{C} \cdot CH_2 \cdot CH(NH_2)COOH$$

The condensation of amino acids involves the formation of peptide links:

$$—\overset{\displaystyle}{\underset{\displaystyle H}{N}}—\overset{\displaystyle}{\underset{\displaystyle O}{C}}—$$

These are formed by a reaction between the amino (basic) group of one acid with the carboxyl (acidic) group of the other. The two condensed amino acids are known as a dipeptide.

$$R^1 \cdot CH \cdot COOH \qquad\qquad R^1 CH \cdot COOH$$

As a result of such a union, it will be seen that one of the amino acids has retained its carboxyl group and the other its amino group, so that further similar condensations are possible with the production of long chains of amino acids:

$$
\begin{array}{cccc}
R^1 & R^2 & R^3 & R^4 \\
| & | & | & | \\
CHCO & CHCO & CHCO & CHCOOH \\
| \quad\backslash\!/ & \backslash\!/ & \backslash\!/ & \\
NH_2 \quad NH & NH & NH &
\end{array}
$$

Condensations proceed via dipeptides, polypeptides, peptones and proteoses to proteins as increasing numbers of amino acids are involved. It must be emphasized that whereas the hydrolysis of proteins can easily be demonstrated to take place in the reverse direction, there is no evidence to suggest that the *in vivo* synthesis of proteins involves the intermediate stages of polypeptides, peptones, etc. (see page 111).

The vast number of combinations of amino acids which are possible in protein molecules of large molecular weight (50,000–200,000) and the diversity of their side groups (R^1, R^2, etc.) provides a ready explanation of the high number of extremely specific proteins which exist (see page 9). The long chains of amino acids are in the form of a helix so that there are about 3·5 amino acids in each turn. This helical structure is itself folded in various complex three dimensional patterns which are essential for protein functioning.

It will be noticed that the amino acids contain both basic (amino) and acidic (carboxyl) groups so that they can react with either acids or alkalis, i.e. they are amphoteric.

Depending on the pH of the external solution they will ionize as either cations or anions.

If an electric current is applied, then a movement of the ions, known as electrophoresis, will take place, to either anode or cathode.

If the total effect of the dissociation of all the amino acids in a protein is taken into account, remembering that the acids are of different strengths, the situation is obviously more complex. However, it is obvious that, for any protein, a pH exists at which anions and cations are formed in equal numbers so that electrophoresis will not be apparent. This pH is known as the iso-electric point.

The constituent amino acids of the protein are also able to combine with the cations of the salts of heavy metals (e.g. lead and mercury). When this happens an insoluble compound is formed and the proteins are precipitated.

Reference to the effects of pH and of heavy metals on protein behaviour is made in connexion with the conditions affecting enzyme action.

Enzymes

At their simplest, enzymes may be considered as catalysts produced by living protoplasm, when a catalyst is defined as a substance which can alter the velocity of a chemical reaction without itself undergoing any permanent chemical change. They differ from conventional catalysts in that they are proteins and so are profoundly altered by any change in their environment which affects their chemical or physical properties, and also in their remarkably specific action (which is probably a corollary of their protein structure).

Broadly, enzymes carry out two important functions in the living organism. In the first place, they are able to accelerate or retard chemical reactions. Secondly, by being controlled in their production and by virtue of their specificity, they are able to regulate a number of different reactions at the same time. Thus if we have a series of reactions $A \rightarrow B$, $C \rightarrow D$, $E \rightarrow F$, it is conceivable that under a given set of conditions, it is advantageous, if, say, the formation of B and F is accelerated and of D retarded. If the enzymes were unspecific in their mode of action, secretion of an enzyme would increase the velocity of *all* the reactions, whereas if there are three enzymes, each specific for one reaction, then a mechanism exists for the independent control of each reaction. In addition, control of metabolic sequences is determined by the availability of the appropriate coenzymes.

Among the properties of enzyme controlled reactions may be mentioned:

The effect of temperature. In general increasing the temperature of a chemical reaction results in an increased reaction velocity, and for a temperature rise of ten degrees Centigrade, the velocity is doubled. This is often expressed as a temperature coefficient Q_{10} which is defined as

$$\frac{\text{The rate of a reaction at a temperature of } (t + 10)^\circ \text{ C}}{\text{The rate of the reaction at a temperature of } t^\circ \text{ C}}$$

In the case of enzyme controlled reactions, such a temperature effect is observed up to a temperature of approximately 30° C. At temperatures greater than this, the enzyme is progressively destroyed by a process of protein denaturation. Since denaturation may often have a Q_{10} of several hundreds and the reaction only takes place significantly in the presence of the enzyme, then at temperatures greater than about 30°C there is a marked fall in the rate of the reaction (see fig. 3).

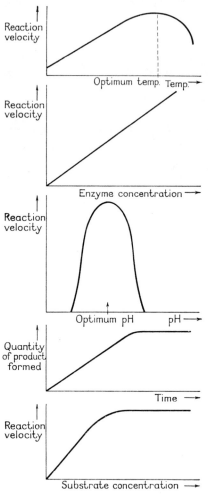

Some of the difficulties in determining optimum temperatures are described on pages 99 and 148.

Confusion can often arise when the results of such experiments are studied. If the reaction is only allowed to proceed for a very short period of time—say two or three minutes—before the reaction velocity is measured, then it may well be found that the rising phase continues to much higher temperatures. This is because there has not been sufficient time for the denaturation process to become significant. It must be emphasised that such measurements give a false impression of the behaviour of the living organism under natural conditions.

The effect of pH is shown in fig. 3. The amphoteric nature of amino acids and of proteins has already been stressed. As a consequence of this it is found that enzymes

Fig. 3. Effect of 1, temperature; 2, enzyme concentrations; 3, pH; 4, time; 5, substrate concentration, on enzyme activity

are usually very restricted in their action by the hydrogen ion concentration of the medium in which they are acting. The graph drawn shows that there is a well-defined optimum in the relationship of enzyme activity and pH although the optimum need not be the same if the reverse reaction is being investigated. Thus, in the case of the enzyme fumarase, the reaction malic acid $\rightarrow$ fumaric acid has a pH optimum of 7·5, but the optimum of the reverse reaction is 6·2.

The effect of increased enzyme concentration. Assuming that there is an excess of substrate available and that other conditions remain constant, there is usually a linear relationship between the concentration of the enzyme and the rate of the reaction.

The effect of increased substrate concentration. Assuming that enzyme concentration and other conditions remain constant, then, up to a definite limiting value, an increased substrate concentration results in an increased rate of reaction, but, beyond the limiting value, there is no further increase in the rate. This can be explained on the assumption that the reaction involves a reaction between the enzyme and its substrate, and is considered later in this chapter.

The effect of protein precipitants. Chemicals which have the property of precipitating proteins are also able to inhibit enzyme controlled reactions. The salts of the heavy metals have already been mentioned. Other inhibitors include trichloracetic acid and cyanide.

The effect of the length of time of the reaction. At first there is a linear relationship between amount of product formed and the length of time the reaction has proceeded, but eventually a stage is reached after which there is no evidence of further enzyme activity even if some substrate is still present. This is often caused by the formation of the equilibrium mixture, although in some cases the products of the reaction may inhibit the action of the enzyme.

Enzyme Specificity. The term 'specificity' refers to the fact that an individual enzyme is able to catalyse only a limited number of reactions, depending on the chemical nature of the substrate.

Four types of specificity can be recognized. If we consider a series of hypothetical molecules, each containing two of the groups A, B or C with a particular type of linkage between them here represented by either — or - - - (e.g. peptide links or glucoside links), then it is possible to distinguish between Low Specific enzymes (Enzyme X) which will catalyse reactions involving compounds with one particular type of linkage (in this case —) but quite independent of the nature of the groups; Group Specific enzymes (Enzyme Y) in which both the linkage and one part of the molecule (in this case A *and* —) must be of a particular type; and finally Absolute Specific enzymes (Enzyme Z), where both groups of the molecule and the linkage are defined (A—B), i.e. they are specific for one particular substrate. This is illustrated in the table below, where a tick signifies that the enzyme will catalyse the reaction and a cross that there is no catalysis.

Compound	Enzyme X (Low)	Enzyme Y (Group)	Enzyme Z (Absolute)
A—B	√	√	√
A—C	√	√	×
B—C	√	×	×
A - - B	×	×	×
A - - C	×	×	×
B - - C	×	×	×

Examples of such specificity are:

	Enzyme	Linkage	Reaction
Low specificity	Lipases	Ester	Fats ⇌ Fatty acids + Glycerol
Group specificity	Trypsin	Peptide	Protein ⇌ Polypeptides
Absolute specificity	Maltase	Glucose 4α Glucoside	Maltose ⇌ 2 Glucose

In addition, a fourth type of specificity exists—Stereochemical specificity. Optical isomerism was referred to in connexion with the carbohydrates. It also exists in many other biological molecules and it is often found that an enzyme is specific for the naturally occurring isomer. For instance most of the amino acids found in nature are laevorotatory and most of the protease enzymes are specific for *l* amino acids.

Enzyme Activation. The protein nature of enzymes has already been emphasized. If a suspension of an enzyme is dialysed, it is found that the residue, containing the protein part, is no longer active, but that if the dialysate is added, activity is restored.

The protein part is called the *apo-enzyme* and the dialysable prosthetic group is the *co-enzyme*. The complete active enzyme, consisting of apo-enzyme and co-enzyme together, is then described as the *holo-enzyme*. The *specificity* of the complete (holo-) enzyme is a consequence of the structure of the apo-enzyme, and the co-enzyme determines the *nature* of the reaction.

In addition to the requirements of a prosthetic group, which is usually a nucleotide, many enzymes also require the presence of inorganic ions for full activity. These are thought to act by forming a link between the enzyme and its substrate (see page 132).

The Mode of Enzyme Action. It is now generally agreed that enzyme action depends on a combination between the enzyme and the substrate molecules. Presumably this results in bringing the reacting molecules much closer together and so accelerating their reaction. The high specificity often shown by enzymes is attributed to highly specialized molecular configurations on their surface which are thought to allow only a very limited number of substances to be adsorbed. This conception is often referred to as the 'lock and key' analogy because of its obvious basic similarity to a mortice lock.

There are three principal lines of evidence in favour of a combination between enzymes and their substrate. The most direct is Keilin's work with the enzyme peroxidase.

This enzyme catalyses the reaction:

$$H_2O_2 + 2[H] \rightarrow 2H_2O$$

and it is only able to proceed if a hydrogen donor is available to supply the necessary hydrogen.

The enzyme has a well-defined absorption spectrum with bands at 498, 548, 583 and 645 mμ. If hydrogen peroxide, *but no hydrogen donor*, is added then a completely different spectrum is found with only two bands at 530·5 mμ and 561 mμ. This can only be interpreted on the basis of a reaction between the enzyme and its substrate. When a suitable hydrogen donor is added the spectrum reverts back to that characteristic of peroxidase.

Also suggesting interaction between enzyme and substrate is the phenomenon of competitive inhibition. As an example we can consider the inhibition of the enzyme succinic dehydrogenase by malonate. Here there is a distinct relationship between the concentration of the substrate and the degree of inhibition so that as the concentration of the substrate *increases*, the percentage of inhibition *decreases*. Such competitive inhibition is attributed to the closely similar molecular structure of succinic and malonic acids, so that both compounds are able to enter into combination with the enzyme. It is thought that the enzyme–inhibitor complex is stable so that at low concentrations of succinate, malonate is able to combine with the enzyme and preclude the succinate from combining, but at high succinate concentration the latter is able to compete successfully with malonate for suitable positions on the enzyme's surface.

Malonic acid   Succinic acid

The third line of evidence, due to the work of Michaelis, will only be considered briefly here. He argued that *if* an enzyme–substrate complex was formed, then the Law of Mass Action could be applied, in which case it should be possible to calculate the relationship between the velocity of the reaction and the concentration of the substrate, and this of course can be correlated with actual experimental results. As a result of such a method of approach it was found that in fact there was a very close correlation between theoretical expectations and actual measurements.

Nomenclature of Enzymes. Enzymes are named by altering the ending of the substrate to *-ase*, e.g. the enzyme catalysing the hydrolysis of sucrose is sucr*ase*, of fats (lipids), *lipase*. Exceptions to this rule are some enzymes which have been well known for many years, particularly in human physiology, such as ptyalin, pepsin, trypsin, etc. *Inactive* forms of an enzyme are distinguished by the ending *-ogen*, e.g. trypsin*ogen*, chymotrypsin*ogen*.

The Classification of Enzymes. The classification of enzymes is based on the type of reaction they catalyse. In this section, they are considered under five groupings:

(i) Hydrolases.   (iv) Isomerases.
(ii) Lyases.   (v) Oxidoreductases.
(iii) Transferases.

Hydrolases catalyse the general reaction

$$AB + H \cdot OH \rightleftharpoons AH + BOH$$

Examples of hydrolases include carbohydrases, proteases and lipases. They play a particularly important part in converting insoluble food storage materials into soluble forms which can then be transported, e.g. in germination and growth. There is no significant evidence to suggest that they play a major part in the reverse direction, viz. in the synthesis of elaborated materials from simpler soluble sources.

Lipases catalyse the hydrolysis of fats into fatty acids and glycerol, e.g.

$$
\begin{array}{l}
CH_2OOC(CH_2)_{14} \cdot CH_3 \\
| \\
CHOOC(CH_2)_{14} \cdot CH_3 + 3HOH \rightleftharpoons 3 \cdot CH_3(CH_2)_{14}COOH + \\
| \\
CH_2OOC(CH_2)_{14} \cdot CH_3 \\
\text{(Tripalmitin)} \qquad\qquad\qquad \text{(Palmitic acid)} +
\end{array}
\qquad
\begin{array}{l}
CH_2OH \\
| \\
CHOH \\
| \\
CH_2OH \\
\text{(Glycerol)}
\end{array}
$$

This catalysis is important in the germination of seeds with an oil food reserve.

In the majority of cases, the enzyme catalysed reaction is reversible, and the effect of the enzyme is to accelerate *the formation of the equilibrium mixture*. It often appears that in fact the reaction proceeds completely in one direction, e.g. the formation of fatty acids and glycerol. This is because some or all of the products formed are soluble (and hence diffusible) and so are translocated from the sphere of the catalysed reaction which is therefore unable to reach equilibrium.

Carbohydrases catalyse the hydrolysis of carbohydrates, e.g.:
Maltase catalyses the reaction

$$C_{12}H_{22}O_{11} + HOH \rightleftharpoons 2C_6H_{12}O_6$$
$$\text{Maltose} \qquad\qquad \text{2 Glucose}$$

Sucrase (or invertase) catalyses the reaction

$$C_{12}H_{22}O_{11} + HOH \rightleftharpoons C_6H_{12}O_6 + C_6H_{12}O_6$$
$$\text{Sucrose} \qquad\qquad \text{Glucose} + \text{Fructose}$$

Amylase (or diastase) catalyses the reaction

$$(C_6H_{10}O_5)_n + \frac{n}{2}HOH \rightleftharpoons \frac{n}{2}C_{12}H_{22}O_{11}$$
$$\text{Starch} \qquad\qquad \text{Maltose}$$

The enzyme amylase is a complex of at least three enzymes—α amylase, β amylase and R enzyme.

α amylase attacks 1:4 linkages situated anywhere in either amylose or amylopectin—it can attack both terminal pairs of glucose units *and* it can attack 1:4 linkages situated inside the chains (i.e. it is both an exo and an endo 1:4 glucosidase).

β amylase also attacks 1:4 linkages but it is only able to break off maltose molecules from the *ends* of the amylose or amylopectin chains (i.e. is only an *exo* 1:4 glucosidase).

R enzyme, unlike the α and β amylases, is without effect on the 1:4 linkages but it is able to attack the 1:6 linkages which form the branch points in amylopectin molecule.

The possible points of enzyme action, using the convention of page 4, are shown above.

Proteolytic enzymes, catalysing the hydrolysis of proteins and their derivatives, were formerly classified into proteases and peptidases:

Hydrolysis by proteases
————————————————————→
Protein ⇌ proteoses ⇌ peptones ⇌ polypeptides

Hydrolysis by peptidases
————————————————————→
Polypeptides ⇌ dipeptides ⇌ amino acids

The plant enzymes were equated with the well-known mammalian digestive enzymes, pepsin, trypsin and erepsin. In the older literature:

Pepsin catalyses the hydrolysis of proteins → polypeptides at pH's *c.* 2–5.

Trypsin catalyses the hydrolysis of proteins → polypeptides, dipeptides and some amino acids at pH's *c.* 7–8.

Erepsin catalyses the hydrolysis of polypeptides → amino acids at pH's *c.* 7–8.

The modern tendency is to replace the terms proteases and peptidases by *Endopeptidases* and *Exopeptidases* respectively.

Endopeptidases are responsible for the hydrolysis of peptide linkages *within* the protein chain and so break it down into smaller molecules.

The various endopeptidases (pepsin, trypsin) show specificity in connexion with the structure of the amino acids adjacent to the peptide linkages.

There are three principal exopeptidases—carboxypeptidase, aminopeptidase and dipeptidase—and together they constitute the enzyme formerly known as erepsin.

Dipeptidase is only able to catalyse the hydrolysis of the peptide link in a dipeptide.

Carboxy- and aminopeptidases attack polypeptides. Carboxypeptidase is specific in its action to a peptide linkage joining a terminal amino acid with a free carboxyl group. Aminopeptidase is also specific for terminal amino acids but these must have a free amino group—i.e. it attacks at the opposite end of the molecule to carboxypeptidase.

Lyases catalyse the general reaction

$$A B \rightleftharpoons A + B$$

Examples of lyases may be considered under the following headings:

(i) Removal of water. Fumarase catalyses the interconversion of malic and fumaric acids

$$
\begin{array}{ccc}
\text{COOH} & & \text{COOH} \\
| & & | \\
\text{CH}_2 & & \text{CH} \\
| & \rightleftharpoons & \| \qquad + \text{H}_2\text{O} \\
\text{CHOH} & & \text{CH} \\
| & & | \\
\text{COOH} & & \text{COOH} \\
\text{Malic acid} & & \text{Fumaric acid}
\end{array}
$$

This is an intermediate stage in the Krebs' cycle, in which pyruvic acid is oxidized aerobically with the consequent release of a large quantity of energy.

(ii) Removal of carbon dioxide. Carboxylase catalyses the conversion of pyruvic acid to acetaldehyde and carbon dioxide. It is possible that this reaction is not reversible.

$$
\begin{array}{ccc}
CH_3 & & CH_3 \\
| & & | \\
CO & \longrightarrow & CHO + CO_2 \\
| & & \\
COOH & &
\end{array}
$$

This reaction is the first stage in which anaerobic respiration differs from aerobic. The acetaldehyde is reduced, in the absence of oxygen to ethyl alcohol.

(iii) Other reactions. Aldolase catalyses the interconversion of fructofuranose diphosphate with dihydroxyacetone phosphate and phosphoglyceraldehyde.

$$
C_6H_{10}O_6 2\,\textcircled{P} \;\rightleftharpoons\;
\begin{array}{c}
CH_2O\,\textcircled{P} \\
| \\
CHOH \\
| \\
CHO
\end{array}
\;+\;
\begin{array}{c}
CH_2O\,\textcircled{P} \\
| \\
CO \\
| \\
CH_2OH
\end{array}
$$

| Fructofuranose diphosphate | Phospho-glyceraldehyde | Dihydroxyacetone phosphate |

This is an important reaction in respiration. It is the stage where the six-carbon compound is split into two three-carbon compounds. *Transferases* catalyse the general reaction

$$ AB + C \;\rightleftharpoons\; AC + B $$

i.e. a reaction in which an entire group or radical is transferred from one molecule to another. Examples of such transfers are:

(i) Transphosphorylations, catalysed by enzymes known as either transphosphatases or phosphokinases.
Hexokinase catalyses the transfer of the terminal high energy phosphate group from ATP (see page 25) to a hexose, e.g.

Glucose + ATP $\longrightarrow$ Glucose 6 phosphate + ADP
Fructose + ATP $\longrightarrow$ Fructose 6 phosphate + ADP

These reactions provide a means of raising the free energy level of the respiratory substrate before glycolysis (page 134).

(ii) Transamination involves the transfer of an amino group (NH_2) from one molecule to another without the formation of ammonia, e.g.

Glutamic acid/alanine transaminase.

$$
\begin{array}{c}
\underset{|}{COOH} \\
\underset{|}{CH_2} \\
\underset{|}{CH_2} \\
\underset{|}{CHNH_2} \\
COOH \\
\text{Glutamic} \\
\text{acid}
\end{array}
\;+\;
\begin{array}{c}
\underset{|}{CH_3} \\
\underset{|}{CO} \\
COOH \\
\text{Pyruvic} \\
\text{acid}
\end{array}
\;\rightleftharpoons\;
\begin{array}{c}
\underset{|}{COOH} \\
\underset{|}{CH_2} \\
\underset{|}{CH_2} \\
\underset{|}{CO} \\
COOH \\
\alpha \text{ Ketoglutaric} \\
\text{acid}
\end{array}
\;+\;
\begin{array}{c}
\underset{|}{CH_3} \\
\underset{|}{CHNH_2} \\
COOH \\
\text{Alanine}
\end{array}
$$

This type of reaction is of fundamental importance in nitrogen metabolism (chapter 6).

Isomerases. These enzymes are able to catalyse isomeric changes in their substrate and may be represented by the general equation

$$ABC \rightleftharpoons ACB$$

Triose phosphate isomerase catalyses the isomeric conversion between phosphoglyceraldehyde and dihydroxyacetone phosphate

$$
\begin{array}{c}
CH_2O\,\boxed{P} \\
\underset{|}{CHOH} \\
CHO \\
\text{Phospho-} \\
\text{glyceraldehyde}
\end{array}
\;\rightleftharpoons\;
\begin{array}{c}
CH_2O\,\boxed{P} \\
\underset{|}{CO} \\
CH_2OH \\
\text{Dihydroxyacetone} \\
\text{phosphate}
\end{array}
$$

Phosphohexoisomerase catalyses the conversion of glucose 6 phosphate into fructose 6 phosphate.

These two enzymes play important roles in respiration.

Oxidoreductases. There are two types of enzyme in this group.

Dehydrogenases catalyse oxidation by the removal of hydrogen from the substrate. A coenzyme is often involved as a hydrogen acceptor and the general equation may be written as

$$AH_2 \rightleftharpoons A + 2[H]$$

These enzymes are considered in detail in the next section (Biological Oxidations).

Oxidases are able to carry out oxidations by the transference of hydrogen from the substrate to molecular oxygen. Either water or hydrogen peroxide is formed:

$$AH_2 + \tfrac{1}{2}O_2 \rightleftharpoons A + H_2O$$
$$or\; BH_2 + O_2 \rightleftharpoons B + H_2O_2$$

The most important of these enzymes is cytochrome oxidase and this is considered in the next section.

Other examples of oxidases include polyphenol oxidase, responsible for the brown coloration of the cut surfaces of potato tubers, and ascorbic acid oxidase, which in some species plays an important part as a terminal oxidase.

In cases where hydrogen peroxide is formed it never accumulates but is acted on by either peroxidase or catalase.

Peroxidase in the presence of a hydrogen donor catalyses the formation of water. If AH_2 represents a hydrogen donor, then

$$AH_2 + H_2O_2 \longrightarrow A + 2H_2O$$

Catalase results in the production of oxygen

$$H_2O_2 \longrightarrow H_2O + \tfrac{1}{2}O_2$$

Biological Oxidations

At its simplest, a substance is said to be oxidized when it combines chemically with oxygen, and so may be represented in general terms as

$$A + [O] \longrightarrow AO$$

The definition has now been extended to include reactions in which either hydrogen is removed from a compound or in which electrons are lost. It is these types of oxidation which are most frequently found in biological systems. If the loss of hydrogen is represented by the equation

$$AH_2 + B \rightleftharpoons BH_2 + A$$

A is oxidized by a dehydrogenation, B is the hydrogen acceptor (and becomes reduced) and the enzyme is a dehydrogenase.

As examples of such dehydrogenation, the relationships between alcohols, aldehydes and acids will be considered.

$$R{\cdot}CH_2OH + [O] \longrightarrow R{\cdot}CHO + H_2O$$
$$\text{Alcohol} \qquad\qquad \text{Aldehyde}$$
$$R{\cdot}CHO + [O] \longrightarrow R{\cdot}COOH$$
$$\qquad\qquad\qquad \text{Acid}$$

At first sight it appears that the reaction of an aldehyde with an oxidizing agent results in a straightforward addition of oxygen, but it can be shown that the reaction will only take place if water is present: the reaction involves a preliminary hydration and then the hydrate (which in many cases is unstable) is dehydrogenated.

$$R \cdot C \underset{O}{\overset{H}{<}} + H \cdot OH \rightarrow R \cdot C \cdot H \underset{OH}{\overset{OH}{<}}$$

Hydrate

$$R \cdot C \cdot H \underset{OH}{\overset{OH}{<}} \rightarrow R \cdot C \underset{OH}{\overset{O}{<}} + 2[H]$$

The hydrogen then combines with oxygen from the oxidizing agent

$$2[H] + \tfrac{1}{2}O_2 \rightarrow H_2O$$

Therefore, in this case, the oxygen is acting as a hydrogen acceptor.

In biological oxidations it is very rarely that the first hydrogen acceptor is oxygen, although under aerobic conditions it is true to say that it is the *ultimate* acceptor. The first hydrogen acceptor is limited in quantity, so that if an oxidation is to continue it is necessary to find a way of reoxidizing the reduced hydrogen acceptor. Under aerobic conditions this is done by molecular oxygen.

$$AH_2 + B \rightarrow BH_2 + A$$
$$BH_2 + \tfrac{1}{2}O_2 \rightarrow B + H_2O$$

An alternative way of writing this equation, introduced by Professor Baldwin, shows the essential links between the two reactions, and this method will be used frequently in this book.

$$AH_2 \underset{A}{\overset{B}{\underset{\smile}{\times}}} \underset{BH_2}{} \overset{H_2O}{\underset{\tfrac{1}{2}O_2}{\times}}$$

Dehydrogenase

Under anaerobic conditions, the reaction could proceed as far as BH_2, but, unless an alternative method of reoxidizing the BH_2 is found, the oxidation of AH_2 would cease when all the B has been used.

Let us suppose that A is subsequently converted into X and that X can be easily reduced by the addition of hydrogen. If this reduction of X could be linked with the oxidation of BH_2 then the reaction would not be blocked, XH_2 will accumulate and molecular oxygen will not be required.

$$AH_2 \underset{A}{\overset{B}{\times}} \underset{BH_2}{} \overset{XH_2}{\underset{X}{\times}}$$

An important example of such a reaction occurs in anaerobic respiration (see chapter 8), and will be discussed briefly here.

Phosphoglyceraldehyde is oxidized to phosphoglyceric acid and as a consequence a hydrogen acceptor, present in limited quantity,

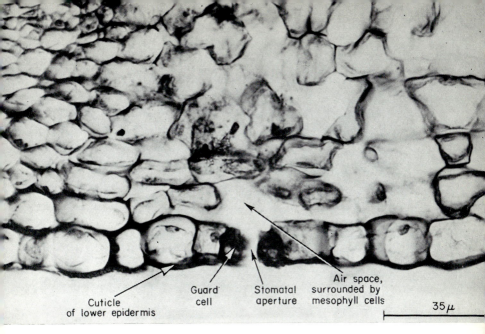

Cuticle
of lower epidermis

Guard
cell

Stomatal
aperture

Air space,
surrounded by
mesophyll cells

35μ

PLATE 1a. Section of Syringa leaf

PLATE 1b. Section of Ranunculus root, to illustrate the
wide cortex of a root

oot hair

Piliferous layer

Cortex showing
intercellular spaces

Xylem

125μ

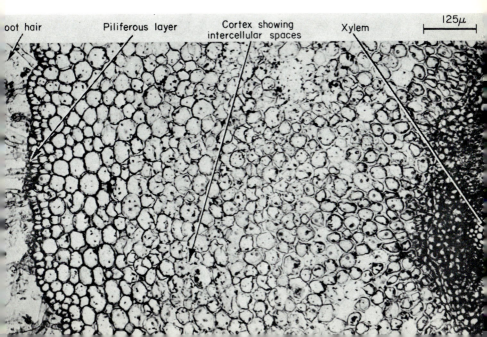

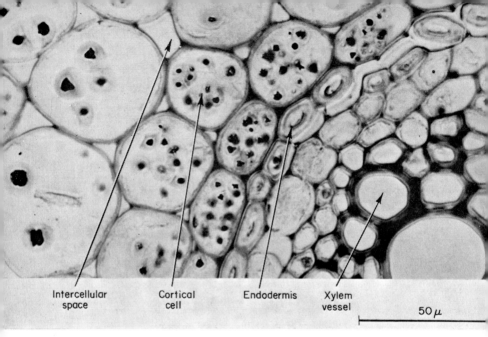

Intercellular   Cortical   Endodermis   Xylem
space   cell   vessel   50 μ

PLATE 2a. Section of Ranunculus root, to show the endodermis and neighbouring cells

PLATE 2b. Section of Iris root

Endodermis   50 μ
with U-shaped thickening   Passage cell

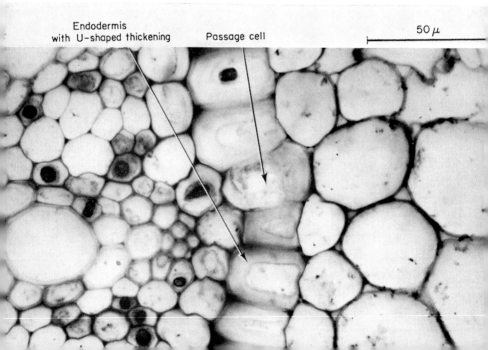

is reduced (i.e. $AH_2 \rightarrow A$ and $B \rightarrow BH_2$). Eventually the phosphoglyceric acid is converted to pyruvic acid which, under anaerobic conditions, loses CO_2 and forms acetaldehyde. This is equivalent to X in the scheme above since the acetaldehyde is reduced to ethyl alcohol utilizing the hydrogen of the reduced hydrogen acceptor which is therefore free to receive more hydrogen from the aldehyde, phosphoglyceraldehyde.

If phosphoglyceraldehyde is written as R·CHO, phosphoglyceric acid as R·COOH, then the essential features can be written in a scheme basically similar to that above:

Glucose
↓

$$R{\cdot}CHO + H_2O \rightarrow \begin{matrix} R{\cdot}CH(OH)_2 \\ R{\cdot}COOH \end{matrix} \Bigg\rangle\!\!\Bigg\langle \begin{matrix} B \\ BH_2 \end{matrix} \Bigg\rangle\!\!\Bigg\langle \begin{matrix} CH_3{\cdot}CH_2OH \\ CH_3{\cdot}CHO \end{matrix}$$

$$\downarrow \qquad\qquad\qquad\qquad \uparrow$$
$$CH_3{\cdot}CO{\cdot}COOH \underline{\qquad\qquad}$$
$$\downarrow$$
$$CO_2$$

(It is interesting to note that a similar principle holds in the anaerobic contraction of mammalian muscle but that here pyruvic acid is reduced to lactic acid—there is no decarboxylation.)

$$\begin{matrix} B \\ BH_2 \end{matrix} \Bigg\rangle\!\!\Bigg\langle \begin{matrix} CH_3CHOH{\cdot}COOH \text{ (Lactic acid)} \\ CH_3COCOOH \end{matrix}$$

The Nature of the Hydrogen Acceptor. There are three principal forms of hydrogen acceptor which can receive hydrogen from a substrate under the action of the appropriate dehydrogenase. These are*

Coenzyme I—Nicotinamide adenine dinucleotide (NAD)
Coenzyme II—Nicotinamide adenine dinucleotide phosphate (NADP) and
Flavo-adenine dinucleotide (FADN)

In addition there exists cytochrome,† which is *only able to receive hydrogen from reduced FADN*. Under the influence of cytochrome oxidase, reduced cytochrome can transfer its hydrogen to molecular

* Coenzyme I was formerly known as Diphosphopyridine nucleotide (DPN) and Coenzyme II was Triphosphopyridine nucleotide (TPN).

† In this book cytochrome is treated as a single substance. It should however be mentioned that several cytochromes exist, e.g. cytochromes *a*, *b*, *c* and *f*. The first three were demonstrated by Keilin, using spectroscopic methods and cytochrome *f* was discovered by Hill in green leaves.

oxygen. It must be emphasized again (see page 19) that cytochrome oxidase is *not* the only terminal oxidase, but there is little doubt that it is the most important.

The dehydrogenase enzymes are specific in their requirements for a hydrogen acceptor, and three distinct categories can be recognized.

Cytochrome specific dehydrogenases are only able to catalyse the transference of hydrogen from the substrate to FADN, i.e.

$$AH_2 \diagdown \diagup FADN \qquad \diagup Cytochrome\ 2H \diagdown \diagup \tfrac{1}{2}O_2$$
$$A \diagup \diagdown FADN\ 2H \diagup \diagdown Cytochrome \qquad \diagdown H_2O$$

Dehydrogenase   Cytochrome
                                   oxidase

Succinic dehydrogenase provides an example of this

$$Cytochrome + \begin{array}{c} COOH \\ | \\ CH_2 \\ | \\ CH_2 \\ | \\ COOH \end{array} \rightleftharpoons \begin{array}{c} COOH \\ | \\ CH \\ || \\ CH \\ | \\ COOH \end{array} + Cytochrome\ 2H$$

Succinic   Fumaric
acid   acid

NAD specific dehydrogenases are only able to catalyse the transference of hydrogen from substrate to coenzyme I, i.e.

$$BH_2 \diagdown \diagup NAD \qquad \diagup FADN\ 2H \diagdown \diagup Cytochrome \qquad \diagup H_2O$$
$$B \diagup \diagdown NAD\ 2H \diagup \diagdown FADN \qquad \diagdown Cytochrome\ 2H \diagup \diagdown \tfrac{1}{2}O_2$$

Dehydrogenase   Cytochrome
                                       oxidase

Triose phosphate dehydrogenase is an example of this type

$$\begin{array}{c} CH_2O\!\!\;\text{(P)} \\ | \\ CHOH \\ | \\ CHO \end{array} + H\!\cdot\!OH \rightleftharpoons \begin{array}{c} CH_2O\!\!\;\text{(P)} \\ | \\ CHOH \\ \quad | \quad OH \\ CH \\ \diagdown OH \end{array}$$

$$NAD + \begin{array}{c} CH_2O\!\!\;\text{(P)} \\ | \\ CHOH \\ \quad | \quad OH \\ CH \\ \diagdown OH \end{array} \rightleftharpoons \begin{array}{c} CH_2O\!\!\;\text{(P)} \\ | \\ CHOH \\ \quad | \quad OH \\ C \\ \diagdown\!\!\! O \end{array} + NAD\ 2H$$

(A phosphorylation is also involved in this reaction.)

NADP specific dehydrogenases are only able to catalyse the transference of hydrogen from the substrate to coenzyme II.

$$\underset{\text{Dehydrogenase}}{\overset{CH_2}{\underset{C}{|}}\times\underset{\text{NADP 2H}}{\overset{NADP}{}}}\times\underset{FADN}{\overset{FADN\ 2H}{}}\times\underset{\text{Cytochrome 2H}}{\overset{Cytochrome}{}}\times\underset{\text{Cytochrome oxidase}}{\overset{H_2O}{\underset{\frac{1}{2}O_2}{}}}$$

An example of a NADP specific dehydrogenase is in triose phosphate dehydrogenase (from chloroplasts). This is dealt with in greater detail in chapter 5.

$$\begin{array}{l}CH_2O\ \textcircled{P}\\ |\\ CHOH\\ |\\ CHO\end{array} + H_2O + NADP \rightleftharpoons \begin{array}{l}CH_2O\ \textcircled{P}\\ |\\ CHOH\\ |\\ COOH\end{array} + NADP\ 2H$$

Oxidation and Electron Transfer is illustrated in the following example. The only observable difference between cytochrome and reduced cytochrome lies in the valency of the iron atom. In cytochrome it is in the ferric form and in the reduced cytochrome it is in the ferrous.

In terms of hydrogen transfer

$$Fe^{3+} + H \longrightarrow Fe^{2+} + H^+$$

and in terms of electrons (e)

$$H \longrightarrow H^+ + e$$
$$Fe^{3+} + e \longrightarrow Fe^{2+}$$

It would appear that this reaction initially takes place in the reduction of FADN, which also has variable valency metals associated with it (e.g. iron, molybdenum, copper, manganese). Considering FADN and cytochrome, but ignoring metals other than iron, the oxidation can be written as

Direction of electron transport

The direction of electron transfer is determined by the redox potentials of the various hydrogen acceptors, but this is beyond the scope of this book.

In the next section of this chapter, the connexion between hydrogen transfer and ATP synthesis will be considered. It is only in that context that reference will be made to the FADN/FADN 2H stage—otherwise only the stages substrate–coenzyme–cytochrome will be mentioned.

The Energy Relationships of Living Cells

An important conception for the understanding of chemical reactions is that of 'free energy'. If by way of illustration we consider an object at the top of an inclined plane, the object will move downwards by virtue of the potential energy which was available for conversion into kinetic energy. Not all the potential energy will be converted—only if we consider the hypothetical case of an infinitely long inclined plane will there be no potential energy left in the system. The important point to notice is that only a portion of the total energy is available for doing work.

Similarly in the case of a chemical reaction, only a portion of the total energy can be liberated for doing useful work—this is the free energy of the system. It must not be confused with the heat of reaction as this only represents a portion of the total amount of free energy. Apart from the free energy, there is the energy which is retained. Entropy gives a measure of the extent of this.

In a chemical reaction there is a tendency for the amount of free energy in the system to decrease, i.e. the reaction is one in which energy is given out (*exergonic*). Such reactions can occur spontaneously if the molecules are in a reactive state (the process of catalysis may be considered as a means of molecular activation). It cannot be too strongly emphasized that the energy given out is given out by virtue of it being 'free' so that it is the reacting compounds which have lost the free energy.

If, on the other hand, an *endergonic* reaction takes place, free energy must be *supplied* to the reactants.

The ultimate source of energy for biological systems is sunlight and this energy is trapped, by photosynthesis, in sugar molecules. Thus photosynthesis is an endergonic reaction and sunlight is the free energy source. This can be lost by respiration which therefore represents a series of exergonic reactions and the free energy lost can be used to drive other synthetic, endergonic, processes.

Adenosine Triphosphate (ATP). This compound occurs in all living cells and consists of the nucleoside adenosine (represented here as A) and three phosphate radicals (—P), i.e.

$$A—P—P—P$$
$$(a)\ (b)\ (c)$$

If the terminal phosphate radical (c) is removed by hydrolysis, there is a loss of free energy amounting to 8–9,000 cal. If the penultimate radical (b) is removed there is a smaller loss of free energy (3,800 cal) but if the last group (a) is split off, only 2,000 cal are liberated.

It can be seen, then, that phosphate radicals can exist at two distinct energy levels—a low level which, incidentally, is characteristic of many simple phosphate esters, and a high level. It is customary to represent the high energy phosphate as ~P and the low energy as —P.

The importance of this is that these two types of phosphate bond* provide an 'energy currency'. When a high energy phosphate is transferred from one compound to another, the bulk of the energy is transferred with it. Also, if low energy phosphate is introduced into a compound which then undergoes energy changes due, for example, to alterations in atomic configurations it is possible for free energy to be transferred to the phosphate, thus converting it into the high energy form.

For the most part it is only ATP which is able to carry out the actual transference of ~P so that the ~P formed from low energy sources is used to regenerate ATP from adenosine diphosphate (ADP) rather than be used directly (in animals the situation is complicated by the existence of another ~P acceptor, creatinine).

The cyclical functioning of ATP is illustrated below.

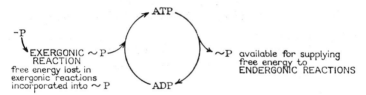

The synthesis of ATP described above is called 'substrate phosphorylation'. Two other methods of forming ATP must also be considered.

Oxidative Phosphorylations. During reactions of the type discussed on page 22, where hydrogen is transferred from a substrate undergoing oxidation via coenzymes, flavoproteins and cytochromes to molecular oxygen, it can be shown that there is always a synthesis of ~P and formation of ATP (i.e. ADP + ~P → ATP).

From measurements of the reduction in free energy, it is unlikely

---

* Although the term 'high energy bond' has been used extensively in the biochemical literature, from the point of view of the physical chemist 'high energy compound' would probably be more acceptable.

that $\sim P$ can be formed in the transfer of hydrogen from substrate to coenzyme I or II, but that it can be formed in the subsequent stages, i.e.

(a) $NAD\ H_2 + ADP + -P + FADN \rightarrow$
$$NAD + ATP + FADN\ H_2$$

(b) $FADN\ H_2 + ADP + -P + Cytochrome \rightarrow$
$$FADN + ATP + Cytochrome\ H_2$$

(c) $Cytochrome\ H_2 + ADP + -P + \frac{1}{2}O_2 \rightarrow$
$$Cytochrome + ATP + H_2O$$

Thus the transference of 2H from either $NAD\ H_2$ *or* $NADP\ H_2$ to molecular oxygen involves the synthesis of at least three molecules of ATP (the actual figure may be greater because the experimental yield is likely to be reduced by the action of the enzyme adenosine triphosphatase).

The existence of up to three intermediate stages in the transference of hydrogen to molecular oxgyen can be looked upon as a method of ensuring that the maximum amount of free energy is fed into the 'energy currency' system of the cell.

The mechanism of oxidative phosphorylation is very imperfectly understood and will not be considered.

Later in this chapter, mention is made of the Krebs' cycle and the facts that

(a) It results in the production of 15 mols. of ATP from the oxidation of one molecule of pyruvic acid.

(b) It can only take place in the presence of molecular oxygen.

These facts can be explained since the majority of the oxidations in the Krebs' cycle involve hydrogen transfer from either NADP or NAD via flavoproteins and cytochromes to molecular oxygen. An ATP 'balance sheet' for the cycle would show

Reaction	No. of ATP molecules	
	Produced	Consumed
Pyruvate → acetate	4	
Acetate + oxaloacetate → citrate		1
Isocitrate → oxalosuccinate	3	
α Ketoglutarate → succinate	4	
Succinate → fumarate	2	
Malate → oxaloacetate	3	
Totals	16	1

∴ Nett production of ATP = 15 mols. per mol. of pyruvate.

If the complete ATP production for aerobic respiration is considered, according to the scheme outlined on page 143.

Reaction	No. of ATP molecules	
	Produced	Consumed
Preliminary hexose phosphorylation		2
2 Triose phosphate → 2 pyruvate	4	
Oxidation of triose (2 mols.)	6	
Krebs' cycle (2 mols. pyruvate)	30	
Totals	40	2

∴ Nett production of ATP per molecule of hexose oxidized = 38 mols.

If the energy value of each $\sim$P is taken as $8 \cdot 9 \times 10^3$ cal, the total energy value is $38 \times 8 \cdot 9 \times 10^3 = 338 \cdot 2 \times 10^3$ cal.

The energy yield for the complete oxidation of one g-mol of glucose is 686×10^3 cal, so that if the energy produced as $\sim$P is considered to be the energy actually available to the plant, the efficiency of the biological oxidation of hexose may be calculated as

$$\frac{338 \cdot 2 \times 10^3 \times 100}{686 \times 10^3} \simeq 48\%$$

Photophosphorylation. In chapter 5 evidence is given to show that ATP synthesis may also occur when chloroplasts are illuminated. The mechanisms involved are very imperfectly understood, but it seems likely that the NADP/FADN/cytochrome pathway of hydrogen transfer is again involved, but the hydrogen is produced by the splitting of water under the influence of light and, under some circumstances, the reduced cytochrome is oxidized by OH from the split water.

The important feature is that solar energy is used to disrupt water molecules. The resulting hydrogen, in a highly reactive state, combines with NADP in an endergonic reaction. A series of exergonic reactions then take place and the progressive losses of free energy are coupled with ATP synthesis. The ATP molecules are in their turn coupled with endergonic stages in sugar synthesis so that solar energy has, in effect, been incorporated into the sugar molecules, where it can be stored until required.

A Metabolic Pathway. A series of important reactions will be considered in order to illustrate some of the essential processes considered in the previous pages. These reactions are fundamental to an understanding of photosynthesis and respiration and the principles underlying them should be thoroughly mastered. In order to simplify the chemistry, emphasis will only be placed on the following aspects:

(*a*) The number of C atoms in a compound.
(*b*) The occurrence of oxidations and reductions.
(*c*) The production and consumption of ATP.

The starting point is the three-carbon compound *phosphoglyceraldehyde* which exists in equilibrium with its keto isomer, dihydroxyacetone phosphate (page 18).

These two compounds can combine together to form *fructose diphosphate* from which polysaccharides can be synthesized via glucose phosphate. The reverse reactions are also possible, and the utilization of ATP in the formation of fructose diphosphate is a *priming reaction*—i.e. an initial raising of the free energy of a compound so that subsequent reactions can take place more rapidly. As stated on page 5, whereas the reaction polysaccharides → hexose phosphate occurs by phosphorylation, it is probable that the reverse reaction of glucose phosphate → polysaccharide involves reactions with uridine triphosphate.

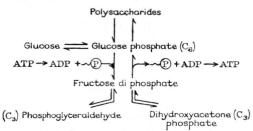

The oxidation of phosphoglyceraldehyde is a key reaction and for the sake of convenience can be considered in two stages.

The actual oxidation follows the pattern outlined on page 20, i.e. the first stage is a hydration, followed by a dehydrogenation. The hydrogen is removed by transference to NAD and eventually to oxygen. However, NAD H_2 reoxidation need not necessarily take place through the cytochrome system—for example it might be coupled with such enzymes as nitrate reductase (page 103).

Inorganic phosphate is also incorporated so that in actual fact *diphosphoglyceric acid* is formed.* It has one *low* and one *high* energy phosphate bond.

* See also page 135.

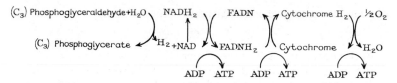

The reduction of phosphoglyceric acid to phosphoglyceraldehyde occurs in photosynthesis. The necessary hydrogen comes from the splitting of water by chlorophyll-absorbed light energy (photolysis: see page 85). The reaction differs from that described above in that NADP is the coenzyme, not NAD.

Diphosphoglyceric acid is converted into *pyruvic* acid. During these reactions, the low energy phosphate is converted into the high energy form so that substrate phosphorylation occurs, i.e. two molecules of ATP are synthesized for each molecule of pyruvic acid formed.

The reactions are shown in fig. 4.

Polysaccharides
via UDP via phosphorylases

Glucose phosphate

ATP ← ADP + ~℗ ⟵ ⟶ ~℗ + ADP ← ATP

Fructose di phosphate

Di hydroxyacetone phosphate

+

H₂O ⟍ ⟋ NADP ⟍ ⟋ Phosphoglyceraldehyde ⟍ ⟋ ½O₂
LIGHT ENERGY | | NAD / 3 ADP
½O₂ ⟋ ⟍ NADP H₂ ⟋ ⟍ Phosphoglycerate ⟍ FADN / 3 ATP
Cytochrome ↘ 3 ATP
↘ H₂O

Di phosphoglyceric acid
⟶ 2 ~℗ + 2 ADP → 2 ATP
PYRUVIC ACID

FIG. 4

Before discussing the fate of pyruvic acid it must be emphasized that whereas the conversion of phosphoglyceric acid to phosphoglyceraldehyde utilizes NADP H_2 when associated with photosynthesis (i.e. when the reaction takes place in the chloroplasts), it can be associated with utilization of NAD H_2 under non-photosynthetic conditions.

The Fate of Pyruvic Acid: The Krebs' Cycle. Under normal aerobic conditions, the pyruvic acid combines with coenzyme A* and carbon

* Coenzyme A is a complex molecule formed from adenosine 3:5 diphosphate, pantothenic acid, and β mercaptoethylamine. It reacts through

dioxide is lost to give a two-carbon compound, acetyl coenzyme A.

$$C_3 \rightarrow C_2 + CO_2$$

Acetyl coenzyme A (C_2) then enters the Krebs' cycle by combination with oxaloacetic acid (C_4) to give iso-citric acid (C_6). By a series of reactions, most of which are oxidations, oxaloacetic acid is reformed, two molecules of CO_2 are evolved and about 14 molecules of ATP are synthesized.

The importance of the cycle is difficult to overestimate. In the diagram the intermediate compounds between iso-citric and α hetoglutaric acid have been omitted. As a consequence of the number of intermediate compounds formed, many side reactions are possible—

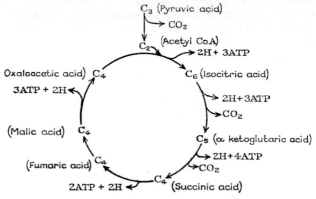

FIG. 5. Summary of Krebs' cycle

e.g. the reaction between α ketoglutaric acid and ammonia to form glutamic acid.

Further points to be noted are—

(a) Since cytochrome *must* be present for the oxidation of succinic acid to fumaric acid (page 22), the cycle can only take place under aerobic conditions (see page 143).

(b) Since the reaction pyruvic acid → acetyl coenzyme A is irreversible, then the cycle as a whole is effectively irreversible despite the large number of reversible reactions which are theoretically possible.

---

the SH group of the latter. In its reaction with pyruvic acid, ATP is involved. Both the high energy phosphate bonds are utilized and their energy becomes incorporated into the S group. Usually coenzyme A is written as CoA·SH and its reaction with pyruvic acid is

$CH_3 \cdot CO \cdot COOH + ATP + CoASH + NAD \rightarrow$

$CH_3 \cdot CO \sim SCoA + AMP + 2P + CO_2 + NAD H_2$
(Acetyl coenzyme A)

2
Diffusion and Osmosis

DIFFUSION

Diffusion can be defined as a movement of molecules from regions of high partial pressure to regions of low partial pressure as a result of their inherent kinetic energy.

The molecules of a fluid are considered to be in a state of continual random movement. Their direction of movement is in a straight line until they collide with another molecule whereupon they will each assume a new course at an angle to the original. Since the molecules of a liquid are more densely arranged than in a gas, then it follows that in a liquid the chances of collision are greater, the path of a molecule more tortuous and the velocity of diffusion is less.

If we imagine a high concentration of molecules of gas X in one part of a container which is otherwise filled uniformly with molecules of gas Y (shown in the diagram as o), it is obvious that the nett direction of movement of molecules of X will be from left to right. As this takes place and the number of molecules occupying the space decreases, Y molecules will tend to enter. The final result will be an even concentration of the two types of molecule over the whole volume.

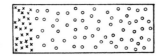

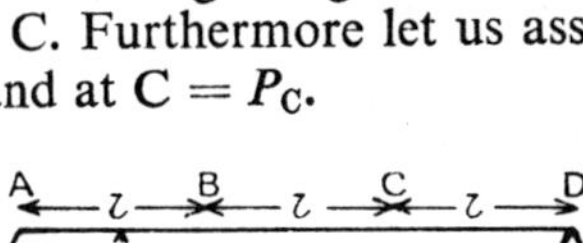

The *rate* of diffusion depends on several factors. To illustrate these let us consider diffusion through length l of a tube of radius r, from, in the diagram, B to C. Furthermore let us assume that the partial pressure at B $= P_B$ and at C $= P_C$.

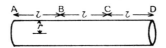

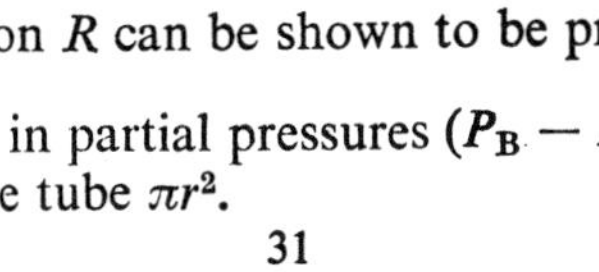

The rate of diffusion R can be shown to be proportional to

(*a*) The difference in partial pressures $(P_B - P_C)$.
(*b*) The area of the tube πr^2.

31

(c) The material *through which* diffusion occurs k.*
(d) Inverse of the length over which diffusion occurs $1/l$.

i.e. $$R = \frac{k(P_B - P_C)\pi r^2}{l}$$

When we come to consider the diffusion of molecules through the end of the tube at D the problem becomes more complicated, because whereas those molecules from the centre of the tube will continue to travel on a course parallel to the longitudinal axis of the tube (paths 6 and 7 in the diagram) those at the edges will be able to pass along paths 1, 2, 3, 4, 5, and 8, 9, 10, 11, 12.

The extent to which this edge effect forms a significant part of the whole diffusion pattern will obviously depend on the relative proportions of the circumference and the area.

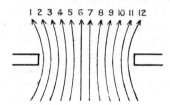

It can easily be shown that the circumference becomes proportionately larger in the case of a *large* number of *small* holes as opposed to a *small* number of *large* holes, but giving the same surface area in each case.

As an example, let us take ten small holes, each of area a sq. cm, and one large hole of area $10a$ sq. cm. Let the radius of each small hole $= r$ cm and of the large hole $= R$ cm.

Then

$$R = \sqrt{\frac{10a}{\pi}}, \quad \therefore \text{ circumference of large hole } = 2\pi\sqrt{\frac{10a}{\pi}}$$

$$r = \sqrt{\frac{a}{\pi}} \quad \therefore \qquad \text{,,} \qquad \text{,, small ,, } = 2\pi\sqrt{\frac{a}{\pi}}$$

Total circumference of the ten small holes $= 20\pi\sqrt{\frac{a}{\pi}}$

$$\therefore \frac{\text{Circumference of 10 small holes}}{\text{Circumference of 1 large hole}} = \frac{10\sqrt{a}}{\sqrt{10a}} \backsimeq 3$$

* k will depend essentially on the resistance the material offers to diffusion. Thus it will be higher for a liquid than a gas. It also varies for different molecules diffusing through the same substance.

Turning for a moment from the tube to consider a multi-perforate septum, as with the stomata of a leaf, the operation of the edge effect depends on the spacing between adjacent stomata. If they are closer together than a distance about ten times their diameter, molecules diffusing from the edge of one pore will meet molecules diffusing from the edge of another pore, thus impeding each other.

As a result of this diffusion pattern at the opening, a series of

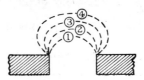

diffusion shells develop. Let us assume that the external partial pressure of the gas is P_ε and that the pressure in the opening is P_D. At a point *immediately* outside the pore, the pressure will be intermediate between P_D and P_ε—say P_1. A little further out it will be intermediate between P_1 and P_ε—say P_2. Thus there are a series of partial pressure shells, each of which can be considered as a contour. In the diagram,

$$P_D > P_1 > P_2 > P_3 > P_4 > P_\varepsilon$$

If we take the case of a molecule diffusing from C to the exterior where the pressure is P_ε, the molecule not only has to travel the distance l but also through the gradually decreasing resistance of the diffusion shells. This has been found to be equivalent to an extra distance $\pi r/2$.

The value for R from C to the exterior (referring to diagram on page 31) would be

$$\frac{k(P_C - P_\varepsilon)\pi r^2}{l + \tfrac{1}{2}\pi r}$$

If we were considering a very thin septum, then l could be ignored and

$$R = 2k(P_C - P_\varepsilon)r$$

i.e. is proportional to the radius, which corresponds with the edge effect, dependent on the circumference.

In the case of stomata l must be considered. From the physical point of view, the stomata may be considered as short tubes of variable aperture, where l may approximate to r.

Importance of Diffusion to the Plant

Diffusion is the principle means by which gaseous exchange takes place. In aquatic acellular plants no structural elaborations are necessary, but in terrestrial plants, quite complex structures are developed for ventilation. In order to explain this, let us take the case of a small hypothetical spherical organism, containing homogeneous protoplasm and with a radius of one unit.

The *volume* will be an index of the amount of metabolizing protoplasm and so of the consumption and production of O_2 and CO_2.

The *area* is an index of the space available through which diffusion can occur.

The *ratio* $\dfrac{\text{Surface area}}{\text{Volume}}$ is a measure of the amount of protoplasm dependent on one square unit of area for its exchanges with the environment.

The table shows how the ratio changes as the organism grows.

Radius		1	2	3	4	5	6
Area		4π	$4\pi4$	$4\pi9$	$4\pi16$	$4\pi25$	$4\pi36$
Volume		$\frac{4}{3}\pi$	$\frac{4}{3}\pi8$	$\frac{4}{3}\pi27$	$\frac{4}{3}\pi64$	$\frac{4}{3}\pi125$	$\frac{4}{3}\pi216$
$\dfrac{\text{Area}}{\text{Volume}}$	$=\dfrac{4\pi r^2}{\frac{4}{3}\pi r^3}$	3	$\frac{3}{2}$	$\frac{3}{3}$	$\frac{3}{4}$	$\frac{3}{5}$	$\frac{3}{6}$

Radius		7	8	9	10..	..n	
Area		$4\pi49$	$4\pi64$	$4\pi81$	$4\pi100$	$4\pi n^2$	
Volume		$\frac{4}{3}\pi343$	$\frac{4}{3}\pi512$	$\frac{4}{3}\pi729$	$\frac{4}{3}\pi1000$	$\frac{4}{3}\pi n^3$	
$\dfrac{\text{Area}}{\text{Volume}}$	$=\dfrac{4\pi r^2}{\frac{4}{3}\pi r^3}$	$\frac{3}{7}$	$\frac{3}{8}$	$\frac{3}{9}$	$\frac{3}{10}$	$3/n$	

Since the volume increases as the *cube* of the radius but the area only as the *square* there is a steadily increasing quantity of protoplasm to be supplied by unit area of surface.

Furthermore, with increasing radius, the diffusion path to protoplasm in the centre becomes longer with a consequent slowing down of the rate. The result of the surface area/volume relationship and the increasing distances is to impose an upper limit on the size of this type of organism: for further increases it is necessary to have local increases in surface area, a means of internal ventilation and, in the case of land plants, a waterproof covering to minimize water loss by evaporation.

The surface area increase is brought about by the intercellular spaces. In the leaf mesophyll this provides a vast area for gaseous exchange and the air spaces provide a pathway for more rapid diffusion than would be possible through an aqueous medium. Some hydrophytes, with root systems embedded in mud, manage to evade the essentially anaerobic conditions in their substratum by extensive aerenchyma in their stele so that rapid diffusion of photosynthetic oxygen takes place from leaves to roots (Plates 1a and 4b).

The Diffusion of Oxygen and Carbon Dioxide

Oxygen is required by the plant for aerobic respiration. It is obtained from two sources

    (*a*) The external atmosphere.
    (*b*) Photosynthesizing cells.

Carbon dioxide is required as a raw material for photosynthesis. It is obtained from

    (*a*) The external atmosphere.
    (*b*) Respiring cells.
    (*c*) The soil solution, as bicarbonate ions (HCO_3').

Entrance to the plant is via stomata and lenticels, and some oxygen may be absorbed through the piliferous layer. Inside the plant, movement occurs through the intercellular spaces. This accounts for the greatest *distance* of movement, but the greatest *resistance* is found when diffusion is in the liquid phases—the water of the cell wall and through the cytoplasm to mitochondria and chloroplasts.

The O_2:CO_2 Balance. The Gay Lussac equations for photosynthesis and respiration (pages 83 and 133) show that quantitatively the two equations are opposite to each other and that, in each one, the volumes of O_2 and CO_2 involved are the same (i.e. for photosynthesis, CO_2 consumption $= O_2$ production; for respiration, CO_2 production $= O_2$ consumption).

Under conditions of intense illumination, the rate of photosynthesis is about ten times the rate of respiration. With decreasing light intensity the photosynthetic rate drops but the respiratory rate stays constant. Thus eventually a point is reached (the *Compensation Point*) at which the rates of the two processes are equal. With a further decrease in light intensity, the photosynthetic rate will become less than the respiratory.

The gas exchange by the plant can easily be calculated. Assume that the respiratory oxygen consumption $= x$ c.c./hr. Then,

High light intensity

    Photosynthesis $10x$ c.c. CO_2 IN   : $10x$ c.c. O_2 OUT
    Respiration x c.c. CO_2 OUT : x c.c. O_2 IN
    ∴ Nett exchange $9x$ c.c. CO_2 IN and $9x$ c.c. O_2 OUT

Reduced light intensity—compensation point

    Photosynthesis x c.c. CO_2 IN   : x c.c. O_2 OUT
    Respiration x c.c. CO_2 OUT : x c.c. O_2 IN
    ∴ Nett exchange—nil, for both CO_2 and O_2

Low light intensity—rate of photosynthesis 50% rate of respiration

Photosynthesis $0.5x$ c.c. CO_2 IN   : $0.5x$ c.c. O_2 OUT
Respiration x c.c. CO_2 OUT   : x c.c. O_2 IN

∴ Nett exchange $0.5x$ c.c. CO_2 OUT : $0.5x$ c.c. O_2 IN

Obviously in the dark, when there is no photosynthesis, only the respiratory exchange takes place.

It is generally accepted that plants cannot exist at light intensities below the compensation point since under these conditions organic material is being used up at a faster rate than it is produced. Many references to compensation point measurements can be found in the limited literature on algal physiology and here the compensation point is usually determined by the depth of water. There are indications that in some cases *Ulva lactuca* can grow abundantly at depths below its compensation point. This could be explained by assuming that there is a change in the respiration rate, so that it decreases with depth, or that the alga can feed saprophytically. A third possibility is that it is capable of absorbing simple soluble organic materials, such as urea, through the cell walls and utilizing them as a raw material for growth.

The Measurement of Gaseous Exchange. In order to measure gaseous exchange—viz. of O_2 and CO_2—it is usual to employ manometric methods. The Warburg and Barcroft respirometers are expensive and extremely sensitive; for elementary work it is advantageous to construct simplified Barcroft respirometers of the type illustrated in fig. 6.

C and F are standard boiling tubes connected by a manometer E, containing paraffin. Both C and F can be connected with the exterior via the rubber connexions and burette clips D and G. B is a graduated tube (e.g. a 5-ml pipette) connected by rubber tubing to a levelling tube A. Both A and B are filled with paraffin. The material under test is placed in C. Seeds can be attached by a hooked pin; chopped tissues can first be placed in a small muslin bag.

Respiratory exchange (viz. O_2 in and CO_2 out) in the case of green tissues must be carried out in the dark, in order to stop photosynthetic exchanges.

Two sets of apparatus are required, X and Y. Samples, as near as possible equal in weight, of the respiratory material are placed in CX and CY. 10 ml of 10% KOH are placed in C and F of apparatus X. In both sets clips D and G are opened to bring pressures on both sides of the manometer to atmospheric, the levels of the menisci in E are marked and the reading of the meniscus at B is noted. Clips D and G are then closed.

As a result of respiration, O_2 in C is consumed and CO_2 is produced.

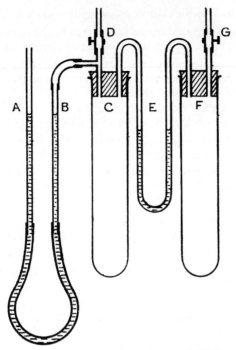

FIG. 6. Modification of Thoday's pattern of a simplified Barcroft apparatus.
(Modified from Thoday, *School Science Review*, 1932, **54**)

In apparatus X, the CO_2 is absorbed by the KOH so that the quantity of gas in C will decrease and this decrease will be due to the volume of O_2 consumed. The meniscus in E will rise towards C. By manipulation of levelling tube A (raising) the meniscus in E can be restored to its original position and the difference between the new and the old readings in B will equal the volume of O_2 consumed.

In apparatus Y there is no absorption of the CO_2, so that any change in volume of C (measured by returning the menisci at E to their original position by the use of the levelling tube) will be due to the *difference* between O_2 consumption and CO_2 production. If the volume changes measured in X and Y are corrected for any difference in weights of tissues, then obviously the CO_2 production can be simply calculated.

If O_2 consumption $= h$ ml
and if O_2 consumption $-$ CO_2 production $= i$ ml
then CO_2 production $= (h - i)$ ml

Photosynthetic exchange (viz. CO_2 in and O_2 out) is complicated by the fact that it is not possible to stop respiration while measuring photosynthesis. It is, therefore, necessary to measure simultaneously

(a) The respiratory exchange (apparatus in the dark).

(b) The total gaseous exchange (apparatus in the light).

The latter is achieved by substituting 1% $NaHCO_3$ for the KOH in tubes C and F of one set of the apparatus since this will maintain a constant CO_2 concentration.

If therefore we have four sets of apparatus, X (light), Y (light), X (dark) and Y (dark), where X (light) contains the $NaHCO_3$,

X (light) will enable the measurement of the volume of O_2 leaving the plant (say j ml).

Y (light) will enable the measurement of the difference between the volume of the CO_2 entering and the O_2 leaving the the plant (say k ml).

In terms of *total* gaseous exchange—

j ml of O_2 leave and $(j - k)$ ml of CO_2 enter

But, the volume of O_2 leaving is less than the volume of O_2 actually formed in photosynthesis because of the respiratory O_2 which is utilized

∴ Volume of O_2 produced in photosynthesis = $\{j + h\}$ ml

Similarly, the volume of CO_2 entering is less than the volume of CO_2 actually used in photosynthesis because of the consumption of the respiratory CO_2.

∴ Volume of CO_2 used in photosynthesis = $\{(j - k) + (h - i)\}$ ml

It will be appreciated that this method is only valid if the rate of respiration is unaffected by the light intensity. Thus the method would probably be invalid if the plant was approaching conditions of carbohydrate starvation.

The Structure and Mode of Action of Stomata

The stomata have already been referred to in connexion with diffusion through pores and the gaseous exchange of the plant. Their structure and mode of action are considered at this point and their role in the control of transpiration is dealt with on pages 58–62.

The Structure of Stomata. This is variable but fig. 7 shows a typical example. The wall next to the pore of each guard cell is considerably thickened by the presence of a secondary layer of cellulose, but the other lateral walls are thin. Each cell has a cytoplasmic lining and a central vacuole containing cell sap. In the cytoplasm is a nucleus and a number of chloroplasts, often poorly developed, although it has been shown that they are capable of photosynthesis by demonstrating their fixation of radioactive carbon (C^{14}).

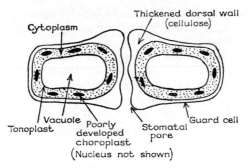

FIG. 7. Basic structure of stomata (vertical section). (See also frontispiece and plate 4*b*)

The Stomatal Mechanism. There is no general agreement as to the stomatal mechanism but most theories assume that a turgor mechanism (page 43) is involved. This is based on the following experimental evidence:

    (*a*) Isolated stomata open when floated on water and close when floated on a hypertonic sugar solution.

    (*b*) The stomata in pieces of isolated epidermis behave in the opposite way—they close when floated on water and open when floated on a hypertonic sugar solution.

    (*c*) Puncturing the larger accessory cells (next to the guard cells) results in the same behaviour as for isolated stomata.

These experiments are interpreted as evidence that stomatal movement is due to the combined water relations of the guard cells and the accessory cells. When the guard cell becomes turgid, the thin walls are extended but the thicker ones, unable to extend appreciably because of their thickening, become slightly concave, thus opening the aperture.

Normally stomatal *opening* is caused by:

The presence of CO_2-free air (in light or dark). Consequently an increase in the external CO_2 level causes closing.

Light. A turgid leaf in the dark, with closed stomata, will show stomatal opening shortly after exposure to light. On returning to the darkness, closing movements start. The light must exceed a minimum intensity in order to be effective, and the intensity also affects the *rate* of opening. The action spectrum (page 87) shows that blue and red light are most effective: it is similar to the action spectrum for photosynthesis, suggesting that the two processes are interlinked and that chlorophyll is the light-absorbing pigment. Light and CO_2 interact in that the higher the light intensity, the greater the CO_2 concentration required for closing.

Water content of the leaf. This must be high for opening to take place.
Temperature. A reduction in temperature from about 30° C often
causes an opening.

A theory to explain stomatal opening must account for a turgor
increase caused by either light or a low external CO_2 concentration,
and decreased turgor by the reverse conditions.

An explanation only involving photosynthesis by the guard cells
(light $\rightarrow$ guard cell photosynthesis $\rightarrow$ accumulation of sugars in
vacuole $\rightarrow$ increased osmotic pressure of sap $\rightarrow$ increased turgor)
is not possible because

(a) The rate of photosynthesis by the guard cells is too low to
account for a sufficient accumulation of sugars.

(b) Photosynthesis would be minimal at low CO_2 concentration,
when turgor would be maximal.

The theory most generally accepted takes account of the following:

(a) Most guard cells contain starch (osmotically inactive because
of its low solubility) in the dark and in the light this starch is
rapidly lost.

(b) Enzymatic conversion of starch $\rightarrow$ sugar occurs at relatively
high pH's.

(c) Enzymatic conversion of sugar $\rightarrow$ starch occurs at relatively
low pH.

(d) An alkaline external pH (e.g. NH_3 vapour) often causes
opening.

The proposed mechanism is

Light $\rightarrow$ high rate of photosynthesis *in mesophyll cells* $\rightarrow$ depletion
of intercellular CO_2
Low concentration of intercellular CO_2 $\rightarrow$ increased pH in guard
cells $\rightarrow$ enzymatic conversion of starch $\rightarrow$ sugar.
Increased sugar content $\rightarrow$ increased osmotic pressure of cell sap $\rightarrow$
increased turgor $\rightarrow$ opening.

Closure in the dark is attributed to an accumulation of respiratory
CO_2 in the intercellular spaces (the CO_2 is no longer removed by
photosynthesis), a drop in pH and therefore a sugar $\rightarrow$ starch reaction with a consequent drop in osmotic pressure and turgor.

This theory is open to the following criticisms:

(a) Some guard cells lack starch (although they usually contain
other polysaccharides).

(b) There is little *quantitative* data concerning the starch $\rightarrow$ sugar
reaction.

(c) Some of the experimental pH's recorded for starch *formation* (using acetate buffers) correspond to the pH's found in *opening* cells.

(*d*) Often closure of stomata at midday takes place without any change in the starch content.

Alternative theories have been put forward, and those suggesting that turgor changes result from active (metabolic) water uptake obviously merit further consideration. As an example the work of Zelitch may be considered. He has suggested that glycollic acid may be involved in stomatal opening and, in support of this, he cites the following—

(*a*) α-hydroxysulphonates (which are competitive inhibitors of glycollic acid oxidase) prevent the opening of stomata in the light.

(*b*) There is a quantitative relationship between the concentration of the inhibitor, the concentration of glycollic acid and the degree of stomatal opening.

(*c*) The inhibition of opening by α-hydroxysulphonates can be reversed by adding glycollic acid.

(*d*) Both stomatal opening and glycollic acid metabolism are inhibited by low oxygen concentrations.

(*e*) Both stomatal opening and glycollic acid metabolism are inhibited to a similar extent by CO_2 concentrations of about 0·5%.

(*f*) The inhibition caused by CO_2 can be reversed by floating the leaf on a solution of glycollic acid.

Zelitch suggests two possible modes of action—either the acid is used as the starting point for sugar synthesis or that it provides a substrate for a process of photophosphorylation and so a means of generating ATP which would provide an energy source for a guard cell pumping mechanism.

OSMOSIS

Osmosis can be defined as 'the movement of molecules of a solvent from a region of high partial pressure of a solvent to a region of low partial pressure when the two regions are separated by a semi-

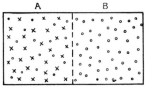

x = sugar molecule

● = water molecule

Fig. 8

permeable membrane' (a membrane permeable to solvent molecules but not to solute molecules).

Fig. 8 represents two equal volumes, A a solution of sugar and B pure water, separated by a semipermeable membrane. If these molecules are all in a state of random motion, then initially only water molecules will collide with the membrane on the B side, but on the A side sugar molecules will also be involved. Any water molecule which reaches a pore in the membrane can pass through to the other side, but sugar molecules cannot pass through because of their size. Since there are more water molecules in B than in A (where they are 'diluted' by the sugar molecules), more water/membrane collisions will occur on the B side and so there will be a nett flow of water from B to A.

If we assume that the water in B is continually replenished, then

(a) A pressure will develop in A. The theoretical maximum value of this pressure, when A is initially filled with pure solute, is known as the *osmotic pressure*.

(b) As water passes into A, the *difference* in partial pressures of water on the two sides of the membrane gradually decreases.

Osmosis is not restricted to the type of system described above where a solution is separated from the *pure* solvent: it will also take place when two solutions are separated by a semipermeable membrane. The direction of movement will still be from regions of high partial pressure of the solvent to a region of low partial pressure (or, from a low concentration (of solute) to a high concentration (of solute)). Osmosis may therefore be considered as a special case of diffusion where movement of some of the molecules is restricted.

The Plant Cell as an Osmotic System

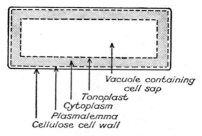

Vacuole containing
cell sap
Tonoplast
Cytoplasm
Plasmalemma
Cellulose cell wall

The main components of the plant cell which are involved in its osmotic relationships are—

The cellulose cell wall: this is a non-living structure with elastic properties. It is *completely* permeable to vacuolar solutes and so plays only an indirect role in the osmotic system.

The plasmalemma and the tonoplast are, respectively, the outer and inner cytoplasmic membranes. They are both partially semipermeable (e.g. sugars and inorganic ions can pass slowly through them). Evidence suggests that the tonoplast may play a more important role than the plasmalemma.

The vacuole, containing the cell sap. The cell sap is an aqueous solution of sugars and mineral salts.

Let us assume that the osmotic pressure of the cell sap is OP_i and that the cell is placed in an external solution with an osmotic pressure OP_ε. Three possibilities exist:

(*a*) $OP_i > OP_\varepsilon$ (*b*) $OP_i < OP_\varepsilon$ (*c*) $OP = OP_\varepsilon$.

When $OP_i > OP_\varepsilon$. Under these conditons (i.e. the external solution is hypotonic) water will enter the vacuole due to the difference in partial pressures of the water inside and outside the cell. The force responsible will be $(OP_i - OP_\varepsilon)$.

As a result of this endosmosis, the volume of the vacuole will increase, the cytoplasm will be forced against the wall and this will be extended due to its elastic properties. The pressure exerted *by* the wall is known as the wall pressure (WP) and the pressure exerted *on* the wall is known as turgor pressure (TP).

Until equilibrium is reached, wall pressure, at any moment, will be slightly less than turgor pressure. When equilibrium is reached, these forces will be numerically equal (but of opposite sign) and the cell is said to be in a state of turgor. At any moment the intake of water will be opposed by the inwardly acting wall pressure, i.e.

Forces causing water to enter the cell $= (OP_i - OP_\varepsilon) + WP$ (where WP has a −ve value).

Suction pressure (SP) is sometimes applied to the forces causing water to enter the cell. It is an inelegant term and *diffusion pressure deficit*, implying a difference in the diffusion (or partial) pressures, is to be preferred.

$$\therefore DPD = (OP_i - OP_\varepsilon) + WP$$

In the special case where the external solution is water,

$$OP_\varepsilon = 0$$
$$\therefore DPD = OP_i + WP$$

This value of the DPD is known as *full* DPD. When the cell is in equilibrium so that there is no further uptake of water,

$$DPD = 0$$
$$\therefore OP_i = WP \quad (= TP)$$

Under these conditions the cell is said to be in a state of *full turgor*. Measurements of *full diffusion pressure deficit* depend on the fact that if cells, or a tissue, are placed in an external solution whose osmotic pressure equals full DPD, then water will neither enter nor leave the cells. But, on the other hand, if the external OP is less than full DPD, water will *enter* the cells and if it is greater, water will *leave* them. Water movement is usually measured by either changes in fresh weight or changes in volume.

The state of turgor is important in the mechanical support of leaves and herbaceous stems. When the plant is well supplied with water, all the living parenchyma cells are fully turgid and so press tightly against each other. The outer layers of cells—collenchyma and epidermis—will consequently be in a state of tension whereas the inner cells will be compressed. This produces a rigid structure.

The presence of these forces can easily be demonstrated by cutting a dandelion scape longitudinally. This releases the compression and tension forces, so that the inner (compressed) cells expand and the outer (extended) cells contract and the scape curls.

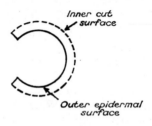

Inner cut surface

Outer epidermal surface

When $OP_i < OP_e$. Under these conditions (i.e. when the external solution is hypertonic) exosmosis will take place so that water will leave the cells, as a result of which

(*a*) The vacuolar volume will decrease.
(*b*) The osmotic pressure of the cell sap will rise.
(*c*) The cell volume will decrease as the elastic wall contracts with decreasing turgor pressure.

A state will therefore be reached at which the cytoplasm *just ceases* to press against the cell wall so that TP = WP = 0.

If at this stage a state of equilibrium is reached, i.e. DPD = 0, then

$$DPD = 0 = (OP_i - OP_e) + WP$$
But
$$WP = 0$$
$$\therefore OP_i = OP_e$$

i.e. at this stage the external solution is isotonic and the cell is in the first stage of plasmolysis—*limiting plasmolysis*

If however the external solution is still hypertonic, then water will continue to be drawn from the vacuole and the second stage of plasmolysis is reached—*incipient plasmolysis*—usually characterized by the fact that the plasmalemma detaches itself from the cell wall at the corners.

Further exosmosis results in greater degrees of plasmolysis. The pattern developed varies according to the solutes present in the external solution and the type of cell. Two common types, convex plasmolysis and concave plasmolysis, are shown diagrammatically in fig. 9.

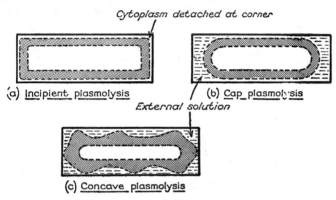

FIG. 9. Diagrammatic representation of types of plasmolysis

Slight plasmolysis is a reversible process, and the cell resumes its normal appearance when placed in a hypotonic solution. In the case of the more advanced forms of plasmolysis, death often results, presumably as a result of damage to the cytoplasm where it is pulled away from the wall. The non-living membranes become completely permeable and there is a fairly rapid exodiffusion of vacuolar solutes from the cell.

On page 43 it was mentioned that the tonoplast was only *partially* semipermeable. As a consequence immediately after plasmolysis has been carried out with such hypertonic solutions as sucrose solution, the osmotic pressure of sucrose solution between the cell wall and the plasmalemma is equal to the osmotic pressure of the cell sap. But, this OP_i is made up of the osmotic activities of sucrose *and* the other vacuolar solutes, so that obviously the concentration of sucrose in the vacuole is less than its concentration in the external solution. A slow inwards diffusion of sucrose occurs with the result of course that the external solution becomes hypotonic with respect to the internal, endosmosis occurs and there is at least a partial recovery from plasmolysis. The sugar mannitol is often used instead of sucrose for plasmolysis studies because it is unable to penetrate the cytoplasmic membranes.

The osmotic pressure of the cell sap can be calculated if the osmotic pressure of an external solution can be found which results in the cell being in equilibrium in a state of limiting plasmolysis, i.e.

$$DPD = 0, \quad WP = 0$$
$$\therefore OP_i = OP_e$$

Unfortunately, both this condition, and that of incipient plasmolysis, where OP_i is slightly less than OP_e, are both extremely difficult to detect by microscopic observations. The method used is to place homogeneous slices of tissue in solutions of known osmotic

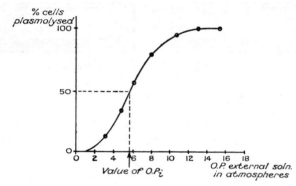

FIG. 10. Typical results for determination of osmotic pressure of cell sap. (Note characteristic sigmoid curve)

pressure for twenty minutes and then to count the number of cells plasmolysed out of a hundred cells examined. (The solutions are chosen so that the most dilute solution does not produce any plasmolysis and the most concentrated produces 100% plasmolysis: also the cells must be counted in a systematic manner—e.g. 'reading' them left to right and line by line.) The results are plotted graphically and by interpolation the osmotic pressure is found of a solution which causes plasmolysis of 50% of the cells. This will give the value for the osmotic pressure of an *average* cell (see fig. 10).

Refinements of the Diffusion Pressure Deficit Equation.

(a) When dealing with cells in a tissue, allowance should be made for pressures caused by surrounding cells. This pressure (Σ) will oppose the entry of water into the cells and will reinforce the wall pressure.

(b) A further allowance must be made for any 'active' water uptake —i.e. any water brought into the cell as a result of energy supplied by metabolic processes (A).

The full equation for the DPD equation is therefore

$$DPD = (OP_i - OP_s) + (WP + \Sigma) + A$$

(Note that both WP and Σ are $-$ve values in that they oppose water uptake.)

Intercellular movement of water

A	B
OP $=$ 10 atm	OP $=$ 15 atm
WP $=$ -2 atm	WP $=$ -8 atm
$\therefore$ DPD $=$ 8 atm	$\therefore$ DPD $=$ 7 atm

In terms of osmotic pressure of cell contents water would be expected to flow from A to B. However, the force governing water entry is the DPD, not the osmotic pressure alone. In the example given the DPD of A (8 atm) is greater than B (7 atm) so that water will move from B to A against the osmotic gradient.

Obviously this simple picture will be complicated by intercellular pressures and also by the relative metabolic activities of the two cells and the consequent occurrence of 'active' water uptake.

Active water uptake. Several lines of evidence suggest the existence of a non-osmotic component in water uptake by cells and tissues. In particular may be mentioned

(*a*) Comparison of plasmolytic and cryoscopic determinations of osmotic pressure.

(*b*) The effects of electrolytes and non-electrolytes on vacuolar volume.

(*c*) The effects of auxins on water uptake.

Comparison of osmotic pressures determined plasmolytically and cryoscopically. The plasmolytic method of determining osmotic pressures has already been discussed. The alternative method is to extract the cell sap by the use of a suitable press and to calculate the osmotic pressure of the expressed fluid by measurement of the depression of freezing point. There are certain difficulties inherent in this method—perhaps the most obvious is a dilution of the sap caused by a filtering action as it is forced through the cytoplasm—but allowances can be made for them. When appropriate corrections have been made, there is often a discrepancy between the two results, *the value determined by the plasmolytic method being higher than that determined by the cryoscopic.*

If it is agreed that the corrected value of the cryoscopic method gives a true value of the osmotic pressure of the cell sap, then it follows that the larger value determined by the plasmolytic method must include a non-osmotic water-absorbing component.

The effect of electrolytes and non-electrolytes on vacuolar volume. The osmotic pressure of the cell sap is determined by a normal plasmolytic method and the cell is then immersed in an isotonic solution of a non-electrolyte (e.g. sucrose) and its vacuolar volume measured (V_S). When the cell is transferred to an *isotonic* solution of an electrolyte (e.g. KCl) it would be expected that the vacuolar volume would remain constant: in fact it changes (to, say, V_e) so that $V_S > V_e$.

Professor Bennet Clark suggested that this was due to an electrical component (metabolically maintained) in water uptake and that the effect of the electrolyte's charged ions was to either decrease or eliminate this component.

Effect of auxins on water uptake. The idea that auxins (see chapter 9) could induce active water absorption was suggested by Reinders who observed

(*a*) Auxin increased the rate of water absorption by potato tissue from distilled water.

(*b*) The process was sensitive to aeration.

(*c*) It was accompanied by an increased loss of dry weight.

(*b*) and (*c*) were interpreted as evidence of an increased rate of respiration.

Further work by Commoner showed that auxin could

(*a*) Prevent exosmosis by potato tissues immersed in hypertonic sucrose.

(*b*) Cause endosmosis by potato tissues immersed in hypertonic sucrose if salts such as potassium fumarate were added.

Since Reinders' work was with *distilled* water, any effect of auxin on salt uptake can be ignored. Presumably the role of potassium fumarate is to increase the respiratory rate by virtue of the role of fumaric acid as an intermediate compound in the Krebs' cycle (page 30).

Water Absorption by the Root as an Example of Osmosis. If a stem is decapitated a few inches above soil level, a watery exudate will often be seen from the cut surface. By attaching a suitable manometer (see fig. 11) a pressure in the order of one or two atmospheres can be demonstrated: this pressure is known as *root pressure* and is commonly considered to be a consequence of the osmotic uptake of water from the soil solution.

The absorbing region of the root is mainly the part containing the root hairs. The extreme tip is of little importance in absorption since the dense cytoplasmic contents of the meristematic cells offer a considerable resistance to water flow. There will also be a high resistance in the older suberized regions, but it is quite possible that some absorption will take place through the lenticels and cracks in the suberized surface. This is likely to be low for a given surface area but the total surface area of suberized root may well result in this absorption playing a significant part in the water economy of the whole plant.

The root hairs have been variously estimated as increasing the surface area of the root by factors of from ×1·5 to ×12. There is no suggestion that root hairs function in any way other than by increasing the surface area.

Mechanism of absorption has been investigated in many experiments which show that for exudation (and therefore root pressure) to develop it is necessary that the osmotic pressure of the exuding sap should be greater than

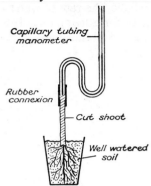

FIG. 11. Determination of root pressure

(a) The DPD of the soil solution.
(b) The DPD of the intervening cortical cells.

Thus variations in the osmotic pressure of the external solution result in corresponding changes in the volume of exudate (decreased external osmotic pressure results in an increased volume of exudate).

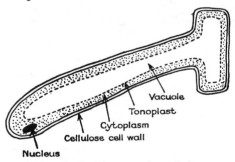

FIG. 12. Structure of root hair

There is no unanimity of opinion as to the pathway traversed by the water and the position of the effective semipermeable membrane. There are three possible routes across the cortex—

(a) Through the cortical cells (in each case including a passage through wall, cytoplasm and vacuole) (fig. 13a).
(b) Through the cellulose walls only (fig. 13b).
(c) Through the cytoplasm (but excluding the tonoplast) and possibly the cellulose walls (fig. 13c).

The water is most likely to pass along the path of least resistance. On this basis, (*a*) can be excluded. In experiments in which decapitated roots are attached to a vacuum pump so that water can be drawn through the roots, it is found that, for a constant force exerted, the rate of water movement is much higher for a dead root than for a living one. This difference can be correlated with the increased permeability of the dead cytoplasm and so implies that a

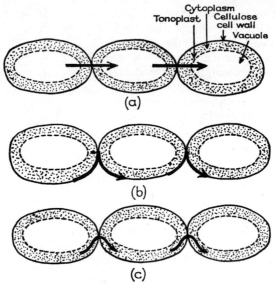

FIG. 13. Possible paths of water movement across root cortex

significant proportion of transverse water movement takes place through the cytoplasm (fig. 13*c*).

The cytoplasm of the parenchymatous cells is continuous between adjacent cells via the plasmodesmata so that uninterrupted movement across the cortex is possible.

It has often been suggested that the structure of the endodermis is such that it could function as the cytoplasmic semipermeable membrane. The presence of the casparian strip of ligno-suberin on the upper, lower and radial walls (fig. 14) will almost certainly ensure that water movement *cannot* take place through the walls (→ A) in this region but it is difficult to see how it can avert water movement through the cytoplasm (→ B) and so ensure that it must take place through the tonoplast (→ C).

Judging by the peculiar behaviour of endodermal cells undergoing plasmolysis it is possible that the cytoplasm may be relatively impermeable (as compared with the cytoplasm of the cortex). If this is the case there would be little movement along (→ B) and the bulk of the water would move along (→ C.)

The osmotically active components of the xylem sap are mainly mineral salts and obviously there must be a continuous supply of

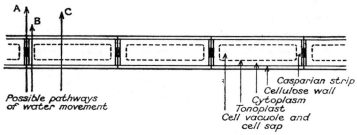

A
B
C

Possible pathways
of water movement

Casparian strip
Cellulose wall
Cytoplasm
Tonoplast
Cell vacuole and
cell sap

FIG. 14. Water movement through the endodermis. (See also plates 1*b*, 2 and 2*b*)

these to the xylem in order to avoid dilution. Almost certainly the salts are secreted into the xylem lumina by an energy-consuming process from the neighbouring parenchyma cells.

As an alternative to active secretion of the salts, Craft and Broyer proposed that there was a passive *leakage* of the salts caused by an accumulation of CO_2 and a deficiency of O_2 in the stele (resulting from the metabolic activities of the root cells). Measurements of oxygen concentrations in the stele do not support this idea.

There is some evidence to suggest that a metabolic pumping of water accounts for at least some of the water entering the xylem. As an example, the work of van Overbeek with tomato plants is particularly illuminating. He first measured the OP of the external solution which *just stopped* exudation. The plants were then transferred to water and the OP of the exudate measured. If the exudation was caused entirely by solutes in the xylem, it should follow that the OP of the external solution just stopping exudation should equal the full (plant in water) OP of the exudate.

In fact he found that the OP of the exudate was *less* than the OP of the external solution, i.e. the external solution was balancing not only an osmotic component but also some other water-providing source.

An unfortunate confusion exists in the use of the word 'active'. It has already been used to describe a non-osmotic water uptake at the cellular level. It is also used to describe *water absorption resulting in root pressure* to distinguish this component of water absorption from the passive component resulting from transpiration (page 63 *et seq.*).

The significance of active water absorption in the life of a plant is difficult to assess. Most measurements show only a small pressure and in general the pressure is minimal when transpiration is maximal.

Probably its main purpose is to provide a subsidiary transport system across the root to the xylem.

Factors affecting the rate of active water absorption are

(*a*) Availability of soil water.
(*b*) Concentration of soil solution.
(*c*) Soil temperature.
(*d*) Soil aeration.

Availability of soil water, after heavy rainfall and a time for drainage, is determined by surface tension forces around the soil particles. This water (representing the field capacity of the soil) is readily accessible to the plant and has an initial DPD of 0·3 to 0·5 atm. As absorption continues, the remaining water becomes increasingly tenaciously held so that, at the permanent wilting point, its DPD is about 15 atm and absorption is very slow. At water contents greater than field capacity, the extra water is also available but it usually results in a displacement of the soil atmosphere with a consequent reduction of the oxygen concentration.

Concentration of soil solutes exerts an effect by increasing the osmotic pressure of the soil solution. In general terms, increasing the concentration of the soil solutes results in a decreased rate of absorption.

Soil temperatures, both low and high, decrease water absorption although it is rare to find temperatures in nature sufficiently high (30–35° C) to limit absorption. The high temperature effect can be correlated with reduced enzyme activity, and the low temperature effect to an increased viscosity of water in addition to a reduced rate of respiration.

Soil aeration is an important factor affecting the rate of active water absorption because, when the oxygen concentration of the soil is low, absorption decreases, presumably due to decreased aerobic metabolism by the root cells and a reduction of salt transference to the xylem.

In cases where the poor aeration is associated with waterlogging of the soil, reduced absorption may also be caused by toxic compounds produced by anaerobic bacteria.

The Donnan Equilibrium. So far emphasis has been placed on the semipermeable nature of the cytoplasmic membrane. A further aspect of membrane phenomena which must be discussed is the effect of having the ions of a salt on both sides of a permeable membrane but with an indiffusible ion (say an anion) on one side of the membrane.

As an example let us consider the salt A^+B' which is completely dissociated. Assume that initially this is present on the right-hand side of a membrane, the concentration of each ion being c_2, and on

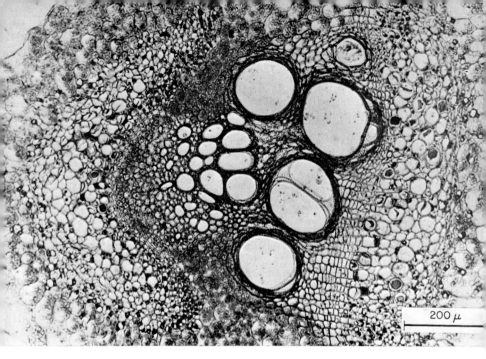

PLATE 3a. Section of Cucurbita stem

PLATE 3b. Longitudinal section of Helianthus stem

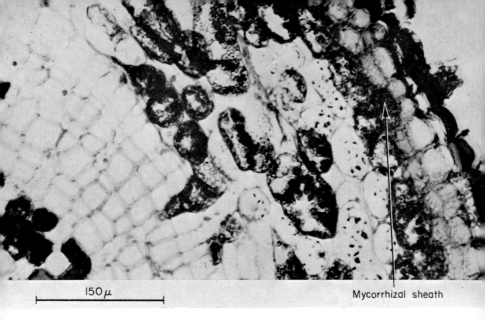

150μ

Mycorrhizal sheath

PLATE 4a. Section of Pinus root

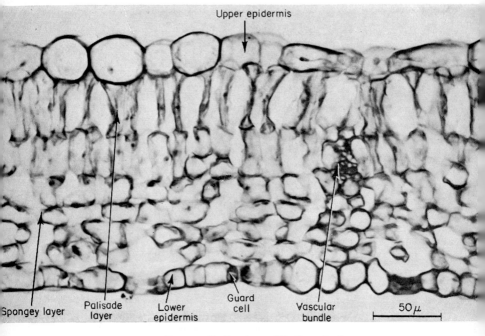

Upper epidermis

Spongey layer | Palisade layer | Lower epidermis | Guard cell | Vascular bundle | 50μ

PLATE 4b. Section of Syringa leaf

the left-hand side is the completely dissociated salt A^+X', each ion having a concentration c_1 and X' being a non-diffusible anion.

Ions		L.H.S.	Permeable membrane	R.H.S.	
	X'	A^+	B'	A^+	B'
Initial conc.	c_1	c_1	0	c_2	c_2
Final conc.	c_1	$(c_1 + x)$	x	$(c_2 - x)$	$(c_2 - x)$

At equilibrium, a certain amount (x) of A^+ and B' ions will have passed across to the left, initially because of the diffusion gradient of B' ions. Electrical neutrality is maintained on both sides of the membrane and of course A^+ and B' ions must pass over in equal concentrations. Furthermore the chemical potentials of the substances present on *both* sides of the membrane (viz. A^+ and B') must be equal. The chemical potentials of an electrolyte may be taken as the sum of the potentials of its ions and this can be shown mathematically to equal the product of their concentrations. Using the conventional square brackets as signifying concentrations,

$$[A^+]_{\text{L.H.S.}} \times [B']_{\text{L.H.S.}} = [A^+]_{\text{R.H.S.}} \times [B']_{\text{R.H.S.}}$$

Substituting
$$(c_1 + x)x = (c_2 - x)(c_2 - x)$$

$$\frac{(c_1 + x)}{(c_2 - x)} = \frac{(c_2 - x)}{x}$$

Clearly $(c_1 + x) > (c_2 - x)$ (this can be proved by assuming first that $(c_1 + x) = (c_2 - x)$ and then that $(c_1 + x) < (c_2 - x)$ and working out each case to a *reductio ad absurdum*.)

From this it follows that

(a) There is a nett movement of ions from R.H.S. to L.H.S.

(b) The number of ions on the L.H.S. is greater than those on the R.H.S. so that an osmotic flow of water will take place to the left.

Of particular importance to the plant is the fact that an actual membrane is not necessary—all that is required is that one ion should be restricted in its movement. This frequently occurs in the cellulose cell walls, with the result that there is an osmotic influx into the walls.

The concept of the Donnan equilibrium is intimately connected with the idea of *free space*—i.e. the part of the cell in which water and solutes can move freely without hindrance by membranes. Structurally this involves the wall and the surface of the cytoplasm. In addition to the salt accumulation resulting from the Donnan equilibrium, there is also evidence to suggest that there is a metabolic uptake of salts. This is discussed on page 79.

3

Transpiration and
the Transpiration Stream

Transpiration can be defined as the loss of water vapour from the
surface of a plant. There are three main places where this can occur:

(a) Stomata (stomatal transpiration), accounting for 80–90% of
the total. Most of the stomata are situated on the leaves, but
they may also occur in the epidermis of herbaceous stems.

(b) Cuticle (cuticular transpiration). The cuticle provides a relatively impermeable covering. If it is thin, up to 20% of the total
transpiration may take place through it, but as its thickness is
increased (e.g. in xerophytes and xeromorphs) the extent of
water vapour loss is reduced.

(c) Lenticels (lenticular transpiration) account for a large proportion of the water loss through the bark, but the water loss from
this source is negligible when compared with the total loss for
the whole plant.

Methods of Measuring Transpiration and Expression of Results

The *intensity* of transpiration relates actual water loss to a given
quantity of plant material over a definite time. The most common
quantity of plant material is in terms of leaf area, although either
fresh or dry weights are sometimes used. Thus results are often expressed as mgm H_2O vapour per sq cm per hour.

Relative transpiration relates water loss from transpiration to water
loss by evaporation from a free water surface (using an atmometer)
under the same external conditions. This ratio, Rate of transpira-
tion/Rate of evaporation (T/E), can provide useful data for indicating the extent to which transpiration is governed by the same conditions as those affecting evaporation.

There are several methods of measuring transpiration rates, none
of which is entirely free from criticism. They may be classified as

(a) Measurement of loss of fresh weight of a plant.
(b) Colorimetric estimation of water loss.
(c) Measurement of amount of water vapour given off.
(d) Potometric methods.

Loss of Fresh Weight. This method assumes that over a period of a few hours the loss in weight by a potted plant with well-watered soil is a measure of its water vapour loss, and that weight changes due to photosynthesis (causing a gain in weight) and respiration (causing a loss in weight) are negligible by comparison. The pot and its contained soil are covered by a polythene bag in order to minimize evaporation from the soil. Over short periods of time this method is extremely reliable, but if the time is too long the results may be complicated by increased water tensions in the soil and also reduced soil aeration.

Colorimetric Estimations. A dry cobalt chloride paper is placed next to the leaf and covered from the air by a microscope slide. The time is measured for a standard colour change from blue (dry) to pink (moist). Although this method has been extensively used, it is doubtful if it measures the rate of transpiration—it is essentially a closed system that is being used and the stomata under the cobalt chloride paper will start to close if the experiment is carried out for more than a few minutes.

Measurement of the Quantity of Water Vapour. This involves passing a stream of air through a small transparent chamber attached to the leaf surface. The air is first passed through a series of U-tubes containing $CaCl_2$ and then, after passing through the chamber, into another series of $CaCl_2$ tubes. The increase in weight of the latter measures the quantity of water vapour given off by the part of the leaf under the chamber. Alternatively, a control experiment may be used and the first set of $CaCl_2$ tubes dispensed with. In this way rates of transpiration at different humidities can be measured. The major objection to this method is that it introduces a 'wind factor'.

Potometric Methods. Such measurements are really a measure of the rates of *absorption*. They give a good indication of factors affecting the rates of transpiration.

Rate of Transpiration. Factors affecting the rate of transpiration can be discussed under two headings—environmental (external) factors and internal factors (features of the plant).

External factors (unless stated to the contrary it is assumed that the stomata are fully open and that there is an abundant supply of water):

(i) *The vapour pressure* deficit i.e. the difference between the water vapour pressure (P_i) in the intercellular spaces and the water vapour pressure in the surrounding air (P_s).

The greater $(P_i - P_e)$, the greater the rate of transpiration.

In the leaf the intercellular spaces will normally be fully saturated (since the spaces are bounded by the wet cell walls) and, because the temperature of the leaf is often higher than that of the surrounding air (as a result of the leaf's metabolism), its saturated vapour pressure may well be slightly higher than the S.V.P. of the surrounding air (i.e. it will correspond to a humidity greater than 100% in relation to the external air). This will allow some transpiration to take place when the surrounding air is fully saturated.

The vapour pressure deficit may be considered the result of the interaction of (a) humidity and (b) temperature.

At a *constant temperature*, decreasing the humidity of the air results in an increased value for $(P_i - P_e)$ and consequently an increased rate of transpiration. This is shown in fig. 15 where the vertical

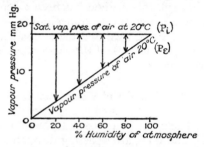

FIG. 15. Relationship between water vapour pressure and humidity at constant temperature (20° C)

double-ended arrows show the vapour pressure deficit for various humidities at 20° C.

At a *constant humidity*, increasing the temperature results in an increased rate of transpiration. This can be considered due to the increased kinetic energy of the molecules and also that at higher temperatures the water-holding capacity of a given volume of air is increased, with a consequent increase in its vapour pressure. This is shown in fig. 16 for a temperature range of 0–50° C and a humidity of 70%.

(ii) *Effect of wind velocity.* On page 33, the idea of diffusion shells over the stomatal apertures was discussed. The total effect of this is to provide a layer of air of a humidity intermediate between that of the air at the stomatal aperture and that of the freely circulating air. The thickness of this layer* varies according to the wind velocity, e.g.

Still air   thickness of relatively stationary air layer = 10 mm
Slow wind   ,,   ,,   ,,   ,,   ,,   ,, = 2·8 mm
Strong wind   ,,   ,,   ,,   ,,   ,,   ,, = 0·4 mm

* Figures of van den Honert.

Since this air layer offers an appreciable resistance to diffusion (see page 33) it is obvious that increases in wind velocity can be expected to increase the transpiration rate until the thickness of the layer is negligible.

Wind also affects the rate of transpiration by causing foliar movement, with a consequent reduction of the humidity in the inter-foliar spaces.

(iii) *Light intensity* causes its main effect by stomatal opening, and

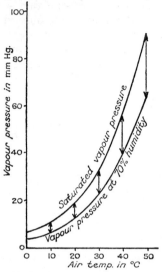

FIG. 16. Relationship between water vapour pressure and temperature at constant humidity (70%)

is discussed under that heading. It has also been reported that increasing light intensity can increase transpiration by causing increased permeability of the protoplasm.

(iv) *The effect of soil water content* on the availability of water to the plant has already been mentioned (page 51). If there is little water available, the resulting tendency for dehydration of the leaf causes stomatal closure and a consequent fall in transpiration.

Internal factors. These can be divided into two types:

(i) *Evidence for a retention of water by the cells*, independent of a stomatal mechanism, has been obtained by a study of relative transpiration (page 54).

If changes in the transpiration rate are accompanied by a *constant* ratio T/E, then it can be reasonably assumed that the factors responsible for the transpiration change are concerned with evaporation. If however the ratio does not remain constant, then it is likely that some factor other than an evaporation factor is concerned.

It has been shown that on bright days, with the stomata fully open, such a change (decrease) in the T/E ratio can be obtained. This is attributed to the process of *incipient drying* of the cellulose walls of the leaf cell. It is shown in a very simplified form in fig. 17.

(ii) There has been considerable controversy as to the effectiveness of *stomatal control in transpiration*. The general view has been that they control transpiration in 'the early stages of an opening movement and the late stages of a closing movement'. Thus Lotfield thought that they only exerted a control if they were less than 50% open and Jeffreys concluded that they were only effective if they were not open more than 2% of their maximum extent.

From the discussion on diffusion through small pores it will be recalled that diffusion resulted in the formation of 'shells' of water vapour and that, if there was any overlapping of these shells, then the

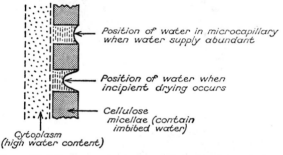

FIG. 17. Diagrammatic representation of incipient drying

'edge effect' was not fully operative. As a general rule it can be stated that stomata must be separated by a distance greater than ten times their diameter to show their maximum effect. It is possible that the observations of Lotfield, Jeffreys and others can be explained by the stomata being close together and only being effectively separated when partially closed.

In most of the work, stomatal aperture was measured by a porometer together with a simultaneous measurement of transpiration rate on some other part of the leaf. An early form of porometer is shown in fig. 18. The cup A is cemented to the underside of the leaf and a suitable manometric fluid is sucked up the vertical tube B by a reduction of pressure at C. The clip at C is then closed and the rate of fall of the fluid in B is measured. This gives an indication of air leakage through the stomata under A and so of their degree of opening.

It has been shown recently that, over a period of time, there is a decrease in the CO_2 concentration in the cup (CO_2 used by photosynthesizing cells above the cup, but not replenished adequately by diffusion) so that the stomata under the cup remain open (page 39) although the stomata used in the transpiration measurements may well be closed. It will be seen that this could explain observations of considerable reductions in transpiration rate associated with little stomatal change.

Obviously it is necessary to re-assess the value of the stomata as transpiration regulators. As a starting point the paper by Professor T. H. van den Honert* is considered. It is included here not only because of its intrinsic value but as an example of a mathematical-physical approach to biological problems.

van den Honert considers water movement through the plant as a *continuous* process (although there is not complete agreement on

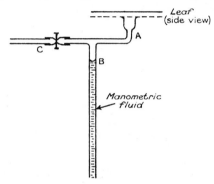

FIG. 18. Simple porometer (after Darwin and Pertz)

this point) and for purposes of simplification he assumes that

(a) There is no active absorption.
(b) Leaf and air temperatures are equal.
(c) Soil water is abundant.
(d) All transpiration is stomatal.
(e) The cohesive theory is correct (see page 64).

He uses the data given on page 56 for the thickness of the stationary air layer and assumes that the air around the leaf has a humidity of 47%.

During its movement through the plant, the water passes through a series of resistances—the root cells (R_R) the stem xylem (R_X) the living leaf cells (R_L) and finally the gaseous stage (R_G) through the intercellular spaces, the stomata and the relatively stationary layer on the leaf surface.

Gradmann pointed out that, if there was a continuous flow, Ohm's

* van den Honert, T. H., 'Water transport in plants as a catenary process', *Disc. Far. Soc.*, No. 3, 1948.

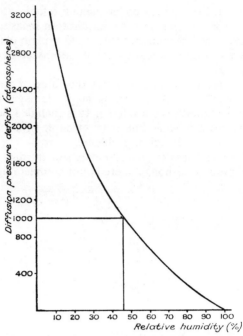

FIG. 19. Relationship between relative humidity and DPD* of the atmospheric moisture at 20° C. (Based on data of Crafts *et al.*, *Water in the Physiology of Plants*, p. 56)

law could be applied to such a system, with a drop in pressure after each resistance—fig. 20.

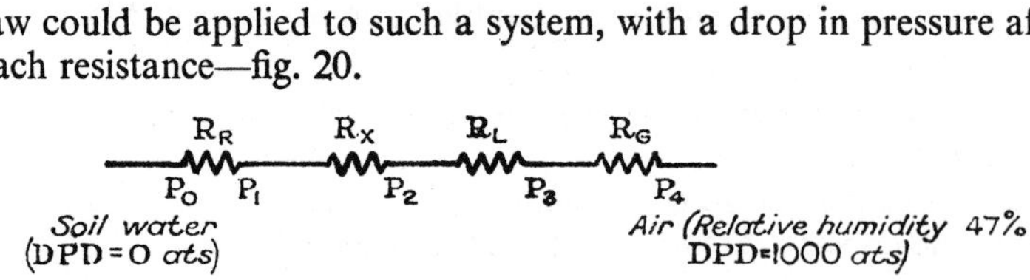

$$R_R \qquad R_X \qquad R_L \qquad R_G$$

$$P_0 \quad P_1 \qquad P_2 \qquad P_3 \qquad P_4$$

Soil water (DPD = 0 ats) Air (Relative humidity 47% DPD = 1000 ats)

FIG. 20. Resistances through the plant

The rate of water movement, dM/dt, will be the same at all points when a steady state is reached and can be represented by the equation

$$\frac{dM}{dt} = \frac{P_1 - P_0}{R_R} = \frac{P_2 - P_1}{R_X} = \frac{P_3 - P_2}{R_L} = \frac{P_4 - P_3}{R_G}$$

If the pressures are considered in terms of DPD (see footnote)

* The DPD of air at a given humidity can be considered simply as the pressure which would have to be applied in order to stop the diffusion (evaporation) of water molecules, from a free water surface, into it.

$P_0 \eqsim$ 0 atm (DPD of the dilute soil soln.)
$P_3 \eqsim$ 50 atm (DPD of the leaf chlorenchyma cells)
$P_4 = 1,000$ atm (DPD of the air, 47% relative humidity)

$$\therefore \frac{P_4 - P_3}{R_G} = \frac{P_3 - P_0}{R_R + R_X + R_L}$$

Substituting

$$\frac{1000 - 50}{R_G} = \frac{50 - 0}{R_R + R_X + R_L}$$

$$\therefore \frac{R_G}{R_R + R_X + R_L} = \frac{950}{50} = \frac{19}{1}$$

In other words the resistance of the gaseous stage is 19 times greater than that of the remaining part of the plant.

To assess the relative resistances of the stomatal pores and the stationary air layer, calculations were based on stomata of the dimensions shown in fig. 21.

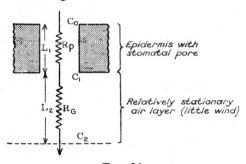

FIG. 21

C_0, C_1, C_2 = water vapour pressures, $l_1 = 0\cdot0005$ cm, $l_2 = 0\cdot5$ cm (relatively little wind). Assume that stomata occupy an area q_1 sq. cm = $\frac{1}{100}$ of the total surface area, q_2 sq. cm D = diffusion constant through air.

$$\text{Rate of diffusion} = \frac{D \times \text{area} \times \text{pressure difference}}{\text{length}}$$

In a continuous flow, rate of diffusion through pore and through the air layer will be equal.

$$\therefore \frac{Dq_1(C_0 - C_1)}{l_1} = \frac{Dq_2(C_1 - C_2)}{l_2}$$

Now

$$\frac{R_P}{R_G} = \frac{(C_0 - C_1)}{(C_1 - C_2)} \quad \text{(from Ohm's law)}$$

and

$$\frac{(C_0 - C_1)}{(C_1 - C_2)} = \frac{q_2 l_1}{l_2 q_1} = \text{(on substitution)} \quad \frac{100 \times 0\cdot0005}{0\cdot5 \times 1}$$

$$= \frac{1}{10}$$

Thus, in relatively still air, the stomata when fully open offer $\frac{1}{11}$ of the *total* resistance of the gaseous stage. In a strong wind (layer thickness = 0·5 mm), R_P and R_G can be shown to be equal. From these calculations it can be deduced that

(1) The main resistance to transpiration is the gaseous stage.
(2) The stomatal control is greatest at high wind velocities.

The crucial point is that stomatal control is situated in the stage offering the greatest resistance (the 'master' reaction) where it will be most effective.

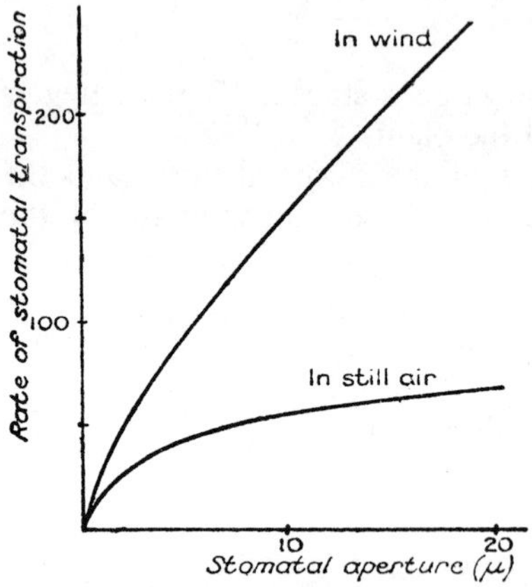

It has already been stressed that this work is largely theoretical and it is interesting to see how it has been substantiated experimentally by Bange.

In his experiments he measured the rates of stomatal transpiration at various stomatal apertures in both still air and wind. He concluded that when the air was still, the stomata exerted a control only if the aperture was small (less than 5μ) but that under windy conditions they exerted a pronounced difference up to much larger apertures (20μ). These results are easily explained on the basis that when there is a wind, there is little resistance offered by the stationary air layer and so stomatal control is maximal, as would be expected from the calculation of van den Honert.

FIG. 22. The relationship between stomatal transpiration and external air movement in a leaf of *Zebrina*. (After Bange, G. G. J., *Acta Bot. Neerl.*, 1953, **2**, 255)

The approach of van den Honert has been subject to some criticism. There have been no actual measurements made of the relative humidity of the leaf atmosphere, so that we only *assume* that it is fully saturated, and there is no evidence to show that a steady state is achieved. The possibility cannot be ruled out that the main resistance to water loss is at the surface of the cells inside the leaf and that incipient drying may be important. Bange's work was with plants which had an ample water supply so that it is unlikely that evaporation from the mesophyll cells would have been a limiting factor in his experiment.

Functions of Transpiration

It has been suggested, at various times, that transpiration functions

(a) Provide the motive power for the absorption and translocation of water and mineral salts.

(b) Provide a cooling mechanism for the plant.

The validity of this is open to question in view of the fact that successful growth can often be achieved under very humid conditions, when transpiration would be very low.

A more commonly held view is that transpiration is a 'necessary evil'. The plant must have a large area for gaseous exchange and it is not possible to have surfaces freely permeable to O_2 and CO_2 but impermeable to water vapour.

The Transpiration Stream

This term refers to the upward movement of water through the xylem, possibly as a result of transpiration.

Evidence that Water Movement takes place in the Xylem. *Ringing experiments*, in which all the tissue external to the xylem is removed from a short length of stem, have little or no effect on the upward movement of water as judged by the presence of unwilted leaves. This suggests that water movement takes place in the xylem, or tissues internal to the xylem, or both.

Movement of dyes (e.g. an aqueous solution of eosin) up a stem when the cut end is placed in the dye results in a staining of the walls of vessels and tracheides, but not of other tissues. Although this indicates that the movement is in the xylem it does not distinguish between movement up the walls (by a wick-like action) or movement in the lumina.

Blockage of the lumina was investigated by Dixon in an attempt to distinguish the relative importance of the walls and lumina. He placed the basal portions of cut shoots in either (A) water (control) or (B) melted gelatine or (C) melted paraffin wax, in each case at 50° C, for forty minutes. After this time the bases were thinly pared (to remove any wax or gelatine in the walls) and all the plants stood in water at 13° C. 15½ hours later the controls' leaves (A) were still fresh but the leaves of B and C had wilted, indicating inadequate water conduction. When the plants were transferred to a concentrated safranin solution it was found, on subsequent microscopic examination, that the walls had stained for a distance of about 20 cm but that the lumina were either completely (C) or almost completely (B) blocked. From this series of experiments it can be concluded that, if the lumina are blocked, there is inadequate water conduction up the stem, even when movement through the walls is still possible.

Is Living Tissue Necessary for the Upward Movement of Water?

Boucherie, in 1840, demonstrated that if a tree was cut across at the base and supplied with a poisonous fluid, it would draw the water up to the top *and* retain the ability to draw up a second sample of the poison. Obviously such a phenomenon is incompatible with any theory requiring the conduction to be dependent on the presence of living cells.* Despite the apparently unambiguous nature of such work, it will be seen later that there is a case to be made out for thinking that the presence of living tissue may be necessary.

The classical explanation of water movement based on purely physical forces is the Cohesion Theory of Dixon and Joly. It explains water movement up the stems of the tallest trees and is discussed here in some detail.

Dixon and Joly's Cohesion Theory. The theory is based on the following premises:

(i) All the water in a plant, but particularly that extending from the root, via the stem xylem to the leaves, is continuous.

(ii) The *cohesive* forces between water molecules, when the water is confined in narrow tubes, is sufficiently high to
  (a) Maintain unbroken columns in the xylem of the highest trees.
  (b) Withstand the added forces caused by friction as the water column moves upwards through the xylem.

(iii) In the mesophyll cells of the leaves, the surface tension forces in the very small microcapillaries of the walls is so high as to balance the oppositely acting forces caused by (ii) *a* and *b*.

(iv) When evaporation occurs from the leaf, the surface tension forces results in water moving up the capillaries and the total effect of this in all the leaf cells is to cause a movement of the entire water mass, i.e. water is dragged through the plant as a result of evaporation from the leaf. The energy for the process is heat energy from the sun.

(v) Particularly in the case of submerged aquatic plants which also show water movement in the xylem, water is *actively secreted* from the leaf cells.

* From a historical point of view it is interesting to note that this is an example of an experiment whose significance was missed. In the 1880's, vitalistic theories (i.e. based on the necessity of living cells for water conduction) were being advanced by Westermaier (1883) and Godlewski (1884) and it was not until 1891 that Strasburger in effect repeated Boucherie's experiments and demonstrated the upward movement of picric acid through the stem of a young oak tree.

There is a difference of opinion as to the path taken by water in the leaf. Evaporation takes place from the walls and water can either (a) be drawn directly from the vacuoles resulting in an increased DPD and so water withdrawal from adjacent cells until eventually it is withdrawn from the xylem; or, (b) move along the *cell walls* of adjacent mesophyll cells until it is withdrawn from the xylem.

It has been suggested that (b) is more likely as it will be a faster process.

A rather different approach to the problem of deciding the pathway of water movement through the mesophyll cells is described by Weatherley.* The two possible pathways can be represented as either a or b in fig. 23.

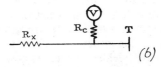

(a)

FIG. 23a. Diagrammatic representation of water movement through the vacuoles

(b)

FIG. 23b. Diagrammatic representation of water movement through the cell walls

Rx = Resistance of movement through the xylem
V = Cell vacuoles
Rc = Resistance of movement through the cytoplasm
T = Effective transpiring surfaces

By constructing appropriate physical models it was found to be possible to obtain characteristic patterns of decreasing absorption when evaporation was stopped from T, and these could be compared with actual measurements on detached leaves attached to a simple potometer when the rate of transpiration was stopped by plunging the leaf into water. The pattern of decreased absorption by the plant corresponded very closely with those found for the model representing water movement through the walls rather than through the vacuoles.

In order to investigate the validity of the cohesive theory it is necessary to examine the three basic assumptions on which it stands:

(a) That the cohesive strength of the xylem sap is adequate.

* Weatherley, P. E., 'The pathway of water movement', in *The Water Relations of Plants*, ed. Rutter and Whitehead, Blackwell, 1963.

(b) That the sap columns are continuous.

(c) That the sap columns are under tension.

Unless *all three* of these conditions hold, the theory is untenable and, furthermore, any experimental conditions which affect the flow of sap by other than purely physical means weaken the theory.

The cohesive strength of xylem sap. Dixon and Joly's original determination was carried out as follows. Samples of xylem sap obtained by centrifuging were introduced into scrupulously clean centrifuge tubes, sealed at one end. The tubes were filled to within a few millimetres of the top before being sealed and were then heated to the 'closing temperature' ($t_C°$ C), i.e. the temperature at which the air bubble just disappeared, so that the tube was completely filled with sap. The tube was then slowly cooled until the 'rupture temperature' ($t_R°$ C) was reached, i.e. when the sap broke with a loud click. Knowing the volume of liquid filling the tube at t_C it was possible to calculate the pressure exerted on the same volume at the lower temperature t_R. The values obtained ranged from 45 to 207 atm.

Subsequent determinations have used a wide variety of methods, and a wide spectrum of results has been obtained. Experiments to determine the tension at which water ruptures in dehiscing fern annuli give values in the order of 200–300 atm whereas purely physical methods have yielded values as low as 0·05 atm. A re-examination of Dixon's work has led to criticisms of his assumption that at the sealing temperature the liquid is at atmospheric pressure —it is likely to be under a pressure of about 100 atm which would result in his values being in the order of six times too large—i.e. his values would range from 8 to 35 atm.

The continuous nature of the sap columns. Many experimental attempts have been made to try and determine whether or not the xylem sap is in continuous columns, but there is no general agreement.

Direct observations of the xylem by carefully peeling back the bark have yielded conflicting results. Some observers report the presence of chains of air bubbles in the vessels and others state that at least some of the xylem vessels are completely filled with sap.

The injection of dyes into the xylem usually results in them moving rapidly in both an upward and downward direction. Although this *could* be interpreted as the result of water movement in two directions it seems more likely to be the result of the xylem contents being both under tension and partially gas-filled.

Freezing of stems in liquid air results in the xylem vessels not being completely filled with ice. However, this type of experiment is rather inconclusive because of the volume changes and tensions caused by different parts of the stem freezing at slightly different times.

Most measurements of transpiration rate and the rate of water movement up the xylem indicate that the water is almost certainly in continuous columns.

The anatomical structure of xylem vessels and tracheides gives few clues. The presence of a torus in bordered pit pairs is mainly confined to some conifers, and in a few cases it is doubtful if the torus would fit sufficiently tightly to act as a seal. It is quite likely that surface tension forces may limit the spread of gas from one xylem vessel to a member of another vertical series, so that the appearance of air bubbles in some conducting tracts does not necessarily mean that they must be present in all.

A different approach to the problem has been adopted, independently, by Preston and Greenidge. In their experiments they show that water movement can still take place when all the xylem elements have been broken and so, presumably, when the continuity of the sap column is interrupted by air. Their experimental technique is very simple—two overlapping saw cuts are made, at intervals, in a tree trunk, ensuring that *all* the xylem elements have been cut. There is relatively little effect on the rates of water movement and transpiration. Greenidge injected dyes into the xylem twelve minutes after the cuts had been made and, apart from some lateral movement and a slight decrease in velocity, the pattern of its distribution was very similar to that of control plants. Preston suggested that water movement takes place through newly differentiated xylem vessels around the edges of the cut.

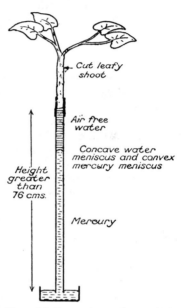

FIG. 24. Apparatus to demonstrate ensions set up by transpiring shoot

Tensions in the xylem are difficult to measure directly. In experiments in which cut leafy branches have been arranged as in fig. 24, it has been found possible to raise mercury to a height greater than 76 cm (although the great difficulty found in setting up such experiments and their sensitivity to movement have prompted some physiologists to suggest that the requirements of the cohesive theory require too delicate a system to be able to withstand the effects of wind on a plant).

Other experiments have involved drawing water through a potometer at a comparable rate to that caused by a plant and measuring the necessary pressure drop required; results of this type of experiment indicate that the xylem sap may be under a tension of as much as 20 atm.

A third line of investigation is to carefully measure the diameter of a stem for various rates of transpiration. As the rate of transpiration increases there is a slight decrease in the stem diameter which would be expected if the xylem contents are under tension.

Of the three conditions which must be fulfilled if the cohesive theory is to be accepted, only measurements of xylem tension give complete support. Much of the evidence for the presence of continuous sap columns of the required tensile strength is far too circumstantial. In view of this it is worth considering the case for vital processes being involved in the ascent of sap.

Recent Vital Theories for the Upward Movement of Xylem Sap. Only a few experiments will be discussed, but they do provide sufficient evidence to make alternatives to the cohesive theory worth considering. The work of Preston and Greenidge has already been mentioned as providing an example of probable movement in air-filled xylem vessels.

Handley's work with cooled stem sections is particularly interesting. He found that if he lowered the temperature of part of a stem to 2° C there was little effect on water movement as judged by the turgid condition of the leaves, but *on lowering the temperature a further 4 deg C* to − *2° C* there was a drastic reduction in the rate of water movement and consequent wilting. Turgor could be regained by raising the temperature to above 2° C. The effect cannot be explained in terms of death of the leaves, since the process is reversible, and there is *no* known physical change which could affect the viscosity of the xylem sap over such a small temperature range. Handley concluded that living cells must be involved in the upward movement of sap, although it is only fair to point out that neither are there any known large vital changes over that temperature range.

Of the other vital theories, those of Priestley, in which he considers that water movement only takes place in the newly differentiating xylem elements and the function of the older vessels is to act as a water store, and of Lund, who suggests that bio-electric potentials in the xylem walls are involved, are particularly worthy of mention.

It is necessary to end on a note of caution. Although many of the background requirements of the cohesion theory are not well founded, it must be remembered that any of the alternative 'vital' theories, no matter how modern they appear, must be capable of providing an explanation of the experimental results of Boucherie and Strasburger!

Passive Water Absorption. Reference has already been made to the continuity of water through the cells walls of the root cells (except perhaps in the endodermis, where the continuity is through the cytoplasm). Upward movement of water in the xylem results in the water being pulled through the cell walls; since the root hair wall is in intimate contact with soil water, this results in an absorption of water. Obviously this component will not be involved in experiments with decapitated shoots.

4

Translocation

It will be evident that there must be a system in the plant whereby the various dissolved substances present are transported from their regions of origin to those regions where they will be utilized or stored.

For example, salts must travel from the roots to the leaves; photosynthetic products from the leaves to storage organs and also to the roots where amino acid synthesis may occur; amino acids from roots to regions of protein synthesis (e.g. the root and stem apices) and stored materials from old leaves into the general body of the plant prior to leaf abscission.

In addition to such well-defined pathways, there is evidence to suggest a more generalized circulation of all dissolved materials.

The Translocation of Organic Materials

The earliest evidence that the phloem is the main tissue involved in the transport of organic materials comes from the results of ringing experiments. These consist essentially of the removal of an outer cylinder of tissue of known depth and cell content from the stem— a popular type of ringing involves the removal of the bark over a few inches length (see fig. 25a).

As an example of early experiments the work of Hanstein (1860) can be considered. He found that if he removed a complete ring of extra cambial tissue from near the base of a cut woody twig which was then placed in water, adventitious roots were formed *above* the ring. If he repeated this experiment with plants which had either an additional inner circle of bundles or bi-collateral bundles or scattered vascular bundles, then adventitious roots were formed mainly *below* the ring. Hanstein concluded that food materials, elaborated by the leaves and necessary for adventitious root formation, are transported downwards in the extra cambial tissue, probably the phloem sieve tubes (see plates 3a and 3b).

Other early investigators found that starch tends to accumulate in the bark above a ring and this, together with the microchemical demonstration of sugars in the sieve tubes, resulted in the general acceptance of the idea that organic materials formed in the leaves move downwards in the sieve tubes.

This concept was seriously challenged by Dixon in 1923. He measured the rate at which starch accumulated in a developing potato tuber and also the cross-sectional area of the phloem in the stolon supplying the tuber. He was therefore able to estimate the velocity at which sugars move in the phloem: this turned out to be about 50 cm per hour, although the rate of *simple diffusion* is only about 0·2 mm per day.

Since he could find no evidence of such rapid rates of movement taking place in the phloem, Dixon suggested that the main channel of movement was in the xylem and he showed that it was possible for dyes to travel both upwards and downwards in the xylem.

Thus for a time it appeared that the phloem might not be associated with the translocation of metabolites, but subsequent work suggests that whereas Dixon's estimates of velocity were very accurate, his conclusions regarding the path of transport were incorrect.

Four lines of evidence substantiate the role of the phloem and they will be considered here in some detail. They are:

(a) The investigations of Mason, Maskell and Phillis on Sea Island cotton.

(b) Schumacher's studies on the effects of eosin and other dyes on translocation.

(c) Experiments of Vernon, Aronoff, Biddulph and others with radioactive tracers.

(d) Weatherley and his co-workers' experiments using aphid mouthparts as very fine micropipettes.

Mason, Maskell and Phillis carried out a very extensive study on Sea Island cotton. They used very large numbers of plants and subjected their results to a close statistical scrutiny. Their experiments are summarized here under three headings.

Diurnal variations of sugar content. The diurnal variation in the total amount of sugars in the leaf led, after a time lag, to a similar variation in their concentration in the bark. This was particularly marked for the variations in sucrose concentration in the part of the bark containing the sieve tubes, but there was very little fluctuation in the low concentration of sugars present in the xylem.

Ringing experiments produced evidence in favour of the view that the downward migration of carbohydrates occurs in the phloem. The complete ringing of a stem below a foliage region, with all the laterals removed below the ring, caused an accumulation of carbohydrates above the ring in the bark, wood and leaves. On the other hand they found that dyes moved across the ringed portions, suggesting that a downward migration of some solutes in the xylem is possible (fig. 25a).

Separation of phloem and xylem. When the phloem and xylem were separated by a ring of paraffin wax paper, downward movement in the phloem continued at normal rates. Also when the ringing technique was modified so that the bark was cut in one position only and levered up from the xylem (instead of being completely removed) sugar transport could be shown to take place along the isolated

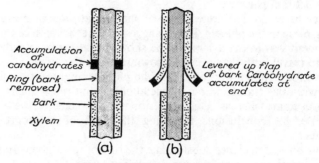

Accumulation of carbohydrates

Ring (bark removed)

Bark

Xylem

Levered up flap of bark. Carbohydrate accumulates at end

(a)   (b)

FIG. 25. Ringing experiments

strips. From these experiments it was concluded that contact between xylem and phloem was not necessary for normal translocation (thus invalidating the criticism that accumulation above a ring was due to damaged or exposed xylem) (figs. 25b and 26).

Although only the results for sugars have been given it must be emphasized that their results also included organic nitrogenous compounds which were similarly affected.

Schumacher introduced eosin into cut regions of leaves. He observed that the dye moved out of the leaf through the phloem and also that it induced the formation of large callus plugs on the sieve

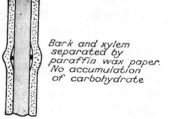

Bark and xylem separated by paraffin wax paper. No accumulation of carbohydrate

FIG. 26. Separation of xylem and bark

plates. He termed this the 'eosin reaction', but he was unable to find any effect on other tissues. The lamina of treated leaves tended to accumulate nitrogenous compounds and this phenomenon was attributed by Schumacher to the plugged sieve plates.

The results of experiments using radioactive tracers also show the importance of the phloem in translocation. Vernon and Aronoff exposed plants to an atmosphere containing $^{14}CO_2$, so that the ^{14}C became incorporated into the photosynthetic end products. By sectioning the stem and placing the sections in contact with X-ray film they were able to show that the radioactivity was confined to the phloem and that the main radioactive component was sucrose.

Weatherley and his co-workers, in a brilliant series of experiments, utilize the fact that aphid mouthparts penetrate into actual sieve tubes. By using the mouthparts as a very fine pipette, they have been able to obtain samples of the contents of single sieve tubes. They have shown that the fluid consists of a strong solution of amino acids and sucrose, and that the concentration of the sucrose varies with the assimilatory activity of the leaves supplying the sieve tube under investigation.

From the four lines of enquiry considered it would appear that

(a) The downward movement of metabolites occurs in the phloem.
(b) The mobile form of carbohydrate in the phloem is sucrose.
(c) Amino acids account for at least some of the organic nitrogenous compounds in the phloem (with asparagine or glutamine as the other principal component—see chapter 6).

It must be emphasized that movement in the phloem can also take place in an upwards direction. Mason *et al.* found that by differential leaf shading it was possible to make darkened leaves *import* sugar from illuminated leaves and that when all the leaves from the top two-thirds of plants were removed, it was possible to demonstrate the accumulation of sugars and nitrogenous compounds in upper regions although they were synthesized in basal regions. In their ringing experiments they also showed that a certain amount of leakage of metabolites from phloem to xylem could occur (just as the reverse leakage of salts from xylem to phloem may take place—see page 81).

It has already been mentioned that Dixon estimated that the velocity of sugar transport is in the region of 50 cm per hour—i.e. that it takes place about 60,000 times faster than normal diffusion. Mason and Maskell estimate that the velocity is 40,000 times faster than normal diffusion, and Biddulph, using tracer methods, suggests a velocity of about 40 cm per hour. There would thus seem to be considerable agreement as to the order of the velocity, although it is perhaps worth mentioning some of the possible objections to the apparently simple method of using tracers. These objections have been dealt with in detail by Canny. Briefly they are

(a) Often no allowance is made of the time taken for the tracer, after entering the plant, to reach the phloem.
(b) If, as is thought, the distribution of the tracer throughout the plant is logarithmic and the form of the logarithmic curve depends on the concentration of the tracer used, then it is possible to show that the distribution of radioactivity depends on the concentration of tracer used. Since it is the distribution of radioactivity which is used to calculate the velocity, then erroneous results can be obtained.

In their work, Mason and Maskell attach great importance to movement along concentration gradients. They found that there is a concentration gradient of hexose sugars from the assimilating cells to the companion cells, a concentration gradient of sucrose in the sieve tubes from leaves to regions of consumption and a final concentration of hexose from the companion cells to the actual cells utilizing the translocated carbohydrate. They therefore suggest that movement takes place along concentration gradients and that the function of the sieve tubes is to effect the interconversion of sucrose and hexose.

More light on the function of the companion cells comes from the work of Wanner, Bauer and others. It was shown that sugar phosphates exist in the mesophyll cells, sucrose but no sugar phosphates or hexoses exist in the phloem and that there is intense phosphatase activity in the companion cells. This work suggests that the leaf gradient is in the form of hexose phosphates and that these are converted into sucrose by the companion cells.

In elementary accounts of photosynthesis, great stress is placed on the production of starch: so much so that in much of the practical work the formation of starch is considered to be indicative of photosynthesis. The reasons for this starch production are

(a) Under conditions conducive to a high rate of photosynthesis, the production of sugars far exceeds the rate at which they can be translocated from the leaf by the sieve tubes.

(b) In most plants sugars cannot be stored as such because of the high osmotic potential which would result in the cells. It is therefore stored as an insoluble compound—starch.

(c) When conditions are poor for photosynthesis, usually during the dark, the starch is converted back into sugar (hexose phosphate), transported to the companion cells, converted into sucrose and then translocated via the sieve tubes to the storage organs.

The theories put forward to try and explain the mechanism of phloem transport fall into two major groups—those which consider that a solution is transported through the sieve tubes and those which consider that only the solutes are moved. The former are all, to some extent at least, related to the original Mass Flow Hypothesis of Münch. The modern forms of such hypotheses consider that the sieve tubes function quite passively and that the solutions are pumped through them by adjacent parenchyma cells. The latter theories often postulate movement along interfaces and in some cases require the supply of metabolic energy to the actual sieve tube elements.

The principle of Münch's original mass flow hypothesis can be explained by reference to a physical model such as that shown in fig. 27. Spheres A and B are both semipermeable membranes. A contains a concentrated sucrose solution and B a dilute solution and the whole is immersed in water. Osmosis tends to cause water movement into both A and B but the greater hydrostatic pressure developed in A results in a nett inflow of water into A, along the connecting tube to B and then out into the water. Dissolved sugar will also move into B until ultimately the sugar concentrations, and hence the osmotic potentials, will be equal in A and B and water movement will cease. In order to convert this into a continuous flow system, it is necessary to add sugars at A and to remove them from B. Fig. 28 shows how this could be achieved in the plant. Sugars are produced in the leaf by photosynthesis and are removed in regions where they are consumed (e.g. in the root where they may be converted to starch for

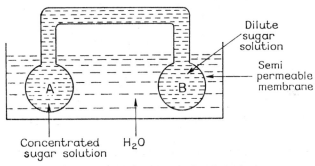

FIG. 27. Principle of mass flow hypothesis

storage) and also water is removed from the leaves by transpiration and is supplied to the roots from the external soil solution. The xylem then forms a conduit from root to leaf (i.e. B → A) and the phloem a conduit from leaf to root (i.e. A → B). In its original form the theory implies a distinct correlation between translocation and the transpiration stream, but in its modern form it is merely held to imply a movement of a solution of metabolites carried passively through the phloem as a result of pumping activities of associated parenchyma cells. It is probably true to say that in some form or other this theory is considered by the majority of physiologists to provide the most likely explanation of the translocation mechanism, and in view of this some of the main criticisms will now be briefly reviewed.

(a) Despite the many cases in which appropriate osmotic gradients have been found (e.g. in the work of Mason and Maskell) it is true to say that a large number of examples have been found in which the gradients exist in the 'wrong' direction.

(b) In the Dioscoraceae, groups of large parenchyma cells in the nodes separate sieve tube elements of the same vertical series.

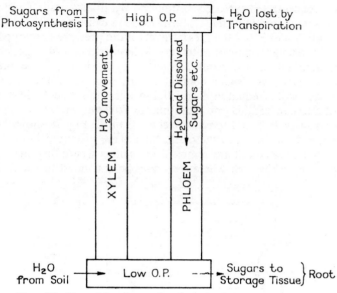

FIG. 28. Essentials of mass flow hypothesis in its original form

If we assume that this family does not have a peculiar mechanism of translocation, then it is difficult to see how mass flow can take place.

(c) The tonoplast is a semipermeable membrane. Early work suggested that this was not so in sieve tubes as it was found to be impossible to plasmolyse sieve tube elements. This was subsequently found to be due to difficulties of technique and there is now no doubt that the tonoplast of sieve tubes is also semi-

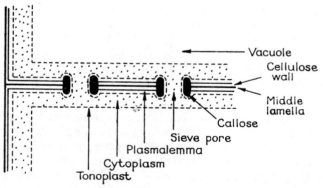

FIG. 29. Structure of sieve plate, based on electron microscope investigations

permeable, and so will provide a high resistance to solute movement from vacuole to cytoplasm.

(d) Electron microscope studies on the structure of the sieve plate suggest that the vacuole is not continuous through the pores, but that these contain strands of cytoplasm bounded by an outer membrane, the plasmalemma (see fig. 29). It is difficult to see how this sieve plate structure can provide other than a high resistance to the movement of solutions.

(e) It is difficult to see how the mass flow hypothesis can explain the simultaneous movement of substances in the phloem in opposite directions at the same time. This objection only holds for two-way movement in the *same* sieve tube and it is difficult to assess the validity of the evidence for such bidirectional transport.

The objections to the mass flow hypothesis can best be summarized by saying that the known structure of the sieve tubes is largely incompatible with the proposed mechanism.

Theories in which only the movement of solutes are considered may be considered to fall into two groups—those in which the movement is considered to take place along an interfacial surface (e.g. the plasmalemma) and those in which an ultra rapid ('activated') diffusion takes place in the cytoplasm.

One of the earliest interfacial theories was advanced by van den Honert in 1932. He postulated that the solute molecules were adsorbed at the interface thus lowering its surface tension and facilitating a rapid spread of the molecules. The necessary motive power was thought to be supplied by the active removal of the molecules from the end of the interfacial path. Unfortunately further investigations suggested that the structure of the cytoplasm would not allow the appropriate surface tension changes to take place. Present-day interfacial hypotheses usually suggest that electro-osmosis may play an important part. Such theories are highly speculative and the underlying mechanisms are well beyond the scope of this book. It must be mentioned, however, that theories which involve the plasmalemma are compatible with the structure of the sieve tube, since the sieve plate provides a region where there is an increase in the total surface area of the outer cytoplasmic membrane and so presumably a decrease in the resistance to movement.

Theories involving 'activated diffusion' imply that energy must somehow be supplied to the molecules in the phloem. There are no very tangible suggestions as to how the activation takes place (it has been suggested that the transport is as phosphates, so that energy could be supplied from ATP. This does not agree with the reports of phosphatase activity in the companion cells (see page 74)), but it is interesting to note that there is an impressive body of evidence to suggest a correlation between the rate of metabolism of the sieve tubes and the rate of translocation. Thus a retardation of the rate of translocation is associated with decreased temperatures, anaerobiosis, potassium cyanide and chloroform, while Kursanov and Turkina, working with sugar beet and *Plantago major*, produce evidence that the respiratory rate of the vascular bundle in petioles is several times greater than that of the surrounding parenchyma. They attribute the respiratory activity of the bundle mainly to phloem, since the xylem is dead and the collenchyma contains little cytoplasm.

There is evidence of a possible connexion between boron and translocation. Gauch and Duggar found that the rate of transport of radioactive sugar in borate-deficient tomato plants could be greatly enhanced by the addition of borate, and they suggest that this may be due to the formation of a borate-sugar complex, possibly in association with the cell membrane. In this case the boron would be associated with electro-osmotic theories, since the complex would function as an anion. Duggar has also produced evidence to suggest that the effect of boron on carbohydrate translocation might be by favouring the formation of hexose in the reaction

$$\text{Glucose 1 PO}_4 \;\rightleftharpoons\; \text{Starch} + \text{PO}_4'''$$

It will be evident to the reader that whereas there is now little doubt that the translocation of organic metabolites takes place in the phloem, there is no proposed mechanism which is free from criticism. The mass flow hypothesis appears to be at variance with sieve tube structure and alternative theories are so speculative as to virtually defy a critical analysis.

The Translocation of Inorganic Material

Before considering the absorption and translocation of mineral salts by the entire plant it will be useful to discuss absorption by individual cells.

It is well established that salt absorption can take place against a concentration gradient (i.e. from a dilute external solution into the relatively concentrated sap in the vacuole) and against an electro-potential gradient. In such cases energy must be supplied to the system and there is considerable evidence to suggest that salt absorption is intimately connected with respiration—in general there is very little absorption under anaerobic conditions and the rate of absorption increases with increases in oxygen concentration. Lundegardh considered that there was a particularly close correlation between the rate of respiration and the absorption of anions and suggested that the total rate of respiration (R_T) was composed of the ground respiration (R_G) (respiration in the absence of salts) and the anion respiration (kA), i.e. $R_T = R_G + kA$.

Many workers consider that the term 'anion respiration' is too restrictive so that the term 'salt respiration' is sometimes used instead.

The mechanism of salt absorption is thought to be associated with the activity of a 'carrier', which is able to combine with, probably, the anion and transport it across the cytoplasm to the vacuole whereupon the anion is discharged into the vacuole and the carrier returns to the periphery and is able to combine with another anion. As a result of the electrical unbalance caused, cations would diffuse into the cells, possibly exchanging with hydrogen ions. In most of the

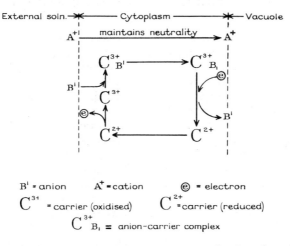

B^1 = anion $\quad$ A^+ = cation $\quad$ ⓔ = electron

C^{3+} = carrier (oxidised) $\quad$ C^{2+} = carrier (reduced)

$C^{3+}{}_{B_1}$ = anion-carrier complex

FIG. 30. A basic carrier mechanism. (Note that this involves the gain of an electron near the vacuole and its loss near the cytoplasmic surface. It is in this electron transfer that respiration is involved. The scheme requires considerable elaboration if it is to explain differential ion accumulation)

schemes it is postulated that the carrier is able to act by virtue of a variable valency component (e.g. $Fe^{2+} \rightleftharpoons Fe^{3+} + e$).

At its simplest such a scheme is shown in fig. 30.

Absorption by the Entire Plants. The majority of the salts are absorbed via the root hairs.* Some of these will be absorbed by the cortical cells of the root but most enter the xylem. As in the case of water absorption the precise pathway for the mineral salts from root hair to xylem is not known. It has been suggested that they are accumulated in the epidermis and then move by diffusion across the

* Mycorrhiza play an important part in the salt absorption of forest trees. This is discussed at the end of the chapter.

intervening cells to the stele, but popular opinion favours the idea that movement across the cortex is either confined to the cell walls or it also moves in the *outer* portion of the cytoplasm which may be very permeable (the Apparent Free Space). There is almost certainly no salt movement via the vacuoles.

If movement is confined to the Apparent Free Spaces (walls and possibly outer cytoplasm) then it is thought that the endodermis may

(*a*) Be the region where salts are accumulated (i.e. the tissue which carries out the metabolic work so that there is a relatively high concentration of salts in the stele).

(*b*) Serve as a barrier so that these salts will not diffuse back into the cortex.

Results of experiments designed to investigate the relationship between salt absorption and the rate of transpiration have often been ambiguous, but Broyer and Hoagland's work has clarified the situation. They found that if there was a low salt content in a plant, reduction of the transpiration rate had little effect on salt absorption but if the salt content was high then a reduced transpiration rate resulted in an approximately parallel reduction in salt absorption.

These results can be explained by assuming that under conditions of low salt concentration, there is relatively little salt accumulated in the stele (by the endodermis) and that these salts can be easily removed in the xylem even when the rate of water movement is low: under these conditions the absorption of salts is limited by the supply available to the endodermis. If however the supply of salts is abundant there will be a considerable accumulation in the stele and, unless these are rapidly removed, they will block the accumulatory mechanism. In this case the rate of absorption is limited by the rate at which the salts can be moved from the stele—that is, on the speed of water movement up the xylem which of course depends on the transpiration rate.

The use of radioactive tracers confirms that salts are carried up in the xylem by the transpiration stream and that they can pass from the xylem into adjacent tissues. Some of these salts can therefore be utilized by parenchymatous cells but others can be shown to reach the phloem and be transported back to the roots.

Salts which reach the leaves may be involved in various metabolic reactions, but those which are not required are, in some cases, re-exported from the leaf in the phloem either back to the roots or to younger leaves, growing points, developing flowers and fruits, etc. Re-export of many minerals from the leaves is also a noticeable feature prior to abscission.

The extent to which this redistribution takes place varies according to the minerals concerned. This has been investigated by Biddulph by spraying the mineral in radioactive form on to the leaf surface and measuring the subsequent spread of radioactivity. Phosphorus moves out of the leaf very rapidly and moves upwards and downwards in the phloem mainly to metabolically active tissues, although leakage into the xylem may occur. Sulphur is similarly transported but calcium shows a distinct contrast; it reaches the leaves via the xylem but it is not mobile in the phloem and is never re-exported.

The Role of Mycorrhiza in Salt Absorption. The term mycorrhiza refers to a symbiotic association between fungi and the roots of higher plants. In this account only the ectotrophic type is considered, these are of particular importance in the case of forest trees. The fungal hyphae form a tightly packed sheath around the roots and also penetrate between the cortical cells to form the Hartig net.

It has been suspected for a long time that the function of the mycorrhizas is to augment the salt absorption of the host plant. Several lines of experimental investigation support this idea, among which may be mentioned:

Comparison of salt absorption between infected and non-infected plants. The experiments of Hatch provide a classical example of this approach to the problem. He grew seedlings of *Pinus strobus* in pots containing a forest-less soil from Wyoming. Half of the pots were inoculated with a mycorrhizal fungus and the other, uninoculated pots kept as controls. He found that the infected plants, when compared with the uninfected plants,

(a) Were a healthy green colour (as opposed to a feeble yellow).
(b) Had a higher dry weight (404·6 gm as against 320·7 gm).
(c) Had significantly higher percentages (dry weight) of nitrogen, phosphate and potassium (see fig. 31).

Direct 'feeding' of mycorrhizal fungi. Melin and his co-workers developed an elegant method by which pine seedlings were grown in sterile sand so that dishes containing mycorrhizal fungi could be placed in contact with them. The hyphae from the cultures grew into the sand and eventually formed an association with the roots. By placing isotopically labelled ions (e.g. PO_4''', NO_3', NH_4^+) in the fungal culture it was possible to follow its translocation along the hyphae and into the root and stem of the seedling, in this way showing that the hyphae were capable of salt absorption and transference.

Experiments with isolated pieces of root have also shown a greater absorption by mycorrhizae over uninfected roots. They also show that the factors affecting salt absorption by mycorrhizae are the same as for normal plants—viz. temperature, oxygen concentration, salt concentration and the presence or absence of metabolic inhibitors.

A further aspect has emerged from this type of experiment, carried out by Harley and his co-workers. They found that with beech mycorrhiza kept in *dilute* phosphate solutions, 90% of the phosphate was absorbed by the fungal sheath and only 10% transferred to the host.

If, subsequently, the pieces were transferred to a phosphate-free solution, some of the stored phosphate was gradually released to the

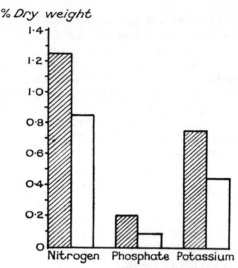

FIG. 31. % Dry weight of N, PO_4''' and K of mycorrhizal (diagonal shading) and non-mycorrhizal plants. (Based on data of Hatch, A. B., *Black Rock Forr. Bull.*, No. 6, 1937)

host. This transfer was sensitive to oxygen concentration (which had to be greater than 3%) and was maximal in air.

Alternatively, if the roots were kept in a phosphate solution, it was found that the stored phosphate was not used but that phosphate was transferred directly from the external solution, via the hyphae, to the host. Curiously, this process was not found to be sensitive to oxygen concentration—possibly diffusion through the 'free space' is involved.

It will be seen that the mycorrhizae not only provide a means of increasing salt absorption by the plant but also can provide a means of salt storage. Plate 4*a* shows the ectotrophic type in a TS of *Pinus* root.

5

Photosynthesis

The term photosynthesis is usually applied to the synthesis of carbohydrates by a plant from simple inorganic sources (CO_2 and H_2O) utilizing light energy which has been absorbed by the chloroplast pigments. As such the process is summarized by the equation

$$6CO_2 + 6H_2O \longrightarrow C_6H_{12}O_6 + 6O_2$$

In its wider sense the term can be extended to include the synthesis of protein and fats: this aspect is considered in chapter 6.

The Supply of Raw Materials for Photosynthesis. Water enters the plant mainly through the root hairs, travels across the cortex and moves upwards in the xylem to the leaves (see chapter 2). Carbon dioxide is obtained from two sources:

(i) The external air. It diffuses through the stomata and the intercellular spaces of the leaves to the wet cell walls of the mesophyll. In its last stage of movement it diffuses through the aqueous phases of the walls and cytoplasm: it is at this point that there is the maximum resistance to diffusion.

(ii) Respiring cells, including the actual photosynthesizing cells. Except in the latter case, where only diffusion from the mitochondria to the chloroplasts is involved, the CO_2 diffuses out of the cells, through the intercellular spaces of the plant and then, as in the case of CO_2 from the air, through the mesophyll walls, and cytoplasm to the chloroplasts.

Stages in Photosynthesis. The equation given above has serious limitations—it implies the rather unlikely occurrence of a reaction between twelve molecules taking place in one stage, and places undue prominence on hexose production.

It is now generally agreed that phosphoglyceraldehyde is the primary photosynthetic product and that the reaction takes place in two main stages:

(*a*) Light stage, for which light is necessary.

(*b*) Dark stage, which is independent of the presence of light.

Evidence for the Existence of Light and Dark Stages. Three pieces of evidence will be considered:

Temperature experiments. Blackman compared the rates of photosynthesis of two groups of plants. One group was kept under a high light intensity but in a low concentration of CO_2 and the other group under a low light intensity but in abundant CO_2. When measurements of the photosynthetic rate were made at different temperatures it was found that in the high light intensity group the $Q_{10} \simeq 2$ and in the low intensity group the $Q_{10} \simeq 1$. Blackman suggested that when the light intensity was high the rate of the overall reaction was limited by the amount of CO_2 whereas when the CO_2 concentration was high then light intensity was the limiting factor.

Warburg's 'Flashing light' experiments. The rate of photosynthesis of a plant kept in continuous light for a certain time was compared with the rate for the plant when supplied with alternate light and dark periods but receiving the *same total amount* of light. The rate of photosynthesis of the latter was found to be significantly greater than that of the plant kept in continuous light.

Both the flashing light and the temperature experiment results can be explained in terms of the reaction sequence

$$A \xrightarrow{\text{Photochemical}} B \xrightarrow{\text{Chemical}} C \quad (C = \text{photosynthetic product})$$

In continuous light of high intensity the reaction $A \rightarrow B$ proceeds at a faster rate than $B \rightarrow C$ so that there is a tendency for some B to accumulate. Under these conditions, increasing the rate of $A \rightarrow B$ will not increase the formation of C because this second reaction is already more than adequately supplied with B.

The effect of intermittent light therefore is to enable $B \rightarrow C$ to proceed in the dark at a time when there is no production of B, as well as for $B \rightarrow C$ to continue in the light. In this way, for a given *total* quantity of light, there is more C produced.

Temperature does not accelerate photochemical reactions. Under conditions of high light intensity, increased temperature accelerates the reaction $B \rightarrow C$ $(Q_{10} \simeq 2)$ at the expense of the excess B.

When there is a *low* light intensity, $A \rightarrow B$ is *slower* than $B \rightarrow C$, so that under these conditions accelerating $B \rightarrow C$ will not affect the overall production of C. Hence the limiting reaction is $A \rightarrow B$ and of course, being photochemical, $Q_{10} \simeq 1$.

Dark pick-up of CO_2 might reasonably be expected to take place if the above explanation is correct. Under conditions of intense illumination but lacking CO_2, 'B' would accumulate and this would react with CO_2 in the dark. This has been demonstrated by subjecting suspensions of algae (e.g. *Chlorella*) to intense illumination in the absence of CO_2 and then rapidly transferring them to a $^{14}CO_2$ containing atmosphere in the dark. The incorporation of this $^{14}CO_2$ could be shown by its radioactivity.

If it is accepted that photosynthesis proceeds by a light and a dark stage, the nature of these two components must be discussed.

The Nature of Light Reaction

Four pieces of experimental work have established quite clearly that the function of the light stage is to produce a 'reducing power'. Hill showed that when isolated chloroplasts of *Stellaria media* were illuminated, oxygen was produced provided that an aqueous leaf extract was also present. Later it was found that the leaf extract could be replaced by a solution of ferric salts (e.g. ferric oxalate) and that the evolution of oxygen was accompanied by a reduction of the ferric ions to ferrous ions. Hill attributed this to the light splitting the water and the resulting hydrogen reducing the ferric ions, i.e.

$$4Fe^{3+} + 2H_2O \xrightarrow{\text{Light, chloroplasts}} 4Fe^{2+} + 4H^+ + O_2$$

Ruben kept *Chlorella* cells *either* in water isotopically enriched with $H_2^{18}O$ but with normal $C^{16}O_2$ (as bicarbonate) *or* in normal water but with isotopically enriched $C^{18}O_2$. In both cases the oxygen produced was tested for $^{18}O_2$ by means of a mass spectrometer and allowances were made for isotopic exchanges. It was found that only when the $^{18}O_2$ was incorporated in the water was there any quantitatively proportional production of gaseous $^{18}O_2$. It must be pointed out that other workers have not obtained such clear-cut results.

Evidence for a reducing role may also be deduced from a study of comparative physiology. If the function of the H_2O is to produce a reducing power, the photosynthetic equation could be summarized as

$$CO_2 + 2H_2O^* \rightarrow (CH_2O) + H_2O + O_2^*$$

(where (CH_2O) represents a monosaccharide and O* shows the fate of the oxygen in water).

In the case of the sulphur bacteria, a similar process takes place but hydrogen sulphide is used as a raw material instead of water and sulphur is deposited instead of oxygen being evolved.

$$CO_2 + 2H_2S \rightarrow (CH_2O) + H_2O + 2S$$

The resemblance between the two equations is obvious, and it is generally accepted that both belong to a class of reactions which can be summarized as

$$CO_2 + 2H_2A \rightarrow (CH_2O) + H_2O + 2A$$

On this view, the 'normal' process of photosynthesis is peculiar in that it utilizes H_2O as the hydrogen donor, with the consequent production of oxygen.

In 1954 Arnon was able to isolate chloroplasts which not only carried out the Hill reaction (see above) but were also capable of ATP production from inorganic phosphate supplied externally. This can be summarized as

$$2Fe^{3+} + H_2O + ADP + P \xrightarrow{\text{Light, chloroplasts}} 2Fe^{2+} + ATP + \tfrac{1}{2}O_2 + 2H^+$$

Later he was able to demonstrate carbohydrate synthesis in isolated chloroplasts and to show that for this a water-soluble chlorophyll-free component of the chloroplast is necessary. It thus became possible to state that the light reaction was associated with the grana of the chloroplasts and the dark reaction with the stroma.

A further aspect of the light stage is the question of light absorption, which is necessary if a photochemical reaction is to take place. In photosynthesis this is achieved by means of the chloroplast pigments—chlorophyll *a*, chlorophyll *b*, carotene and xanthophyll. Some of the properties of these pigments are given below.

The absorption of white light by the chloroplasts is not uniform over the spectrum. The chlorophylls are responsible for the absorption of the red and blue components and the carotenoids are effective in the green (fig. 32).

Pigment	Formula (after Willstatter & Stoll)	Colour	Maximum absorption peaks
Chlorophyll *a*	$C_{55}H_{72}O_5N_4$ Mg	Green	410 and 660 nm
Chlorophyll *b*	$C_{55}H_{70}O_6N_4$ Mg	Green	452 and 642
Carotene	$C_{40}H_{56}$	Orange	449 and 478
Xanthophyll	$C_{40}H_{56}O_2$	Yellow	440 and 490

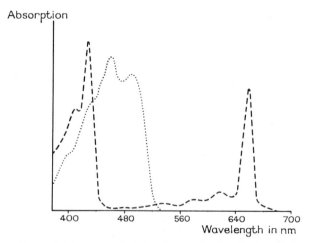

FIG. 32. Absorption spectra of chlorophyll *a* (broken line) and β carotene (dotted line) dissoved in ethyl ether and ether + ethyl alcohol respectively. (From Thomas, *Plant Physiology*, Churchill, 1956)

In order to determine the efficiency of the various wavelengths, it is necessary to study their *action spectra* (which is obtained by measuring the rate of photosynthesis when a plant is illuminated by lights

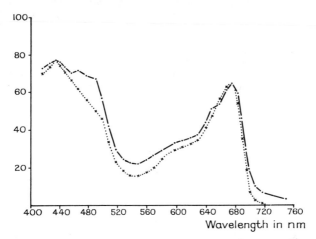

FIG. 33. Absorption spectrum (solid line) and action spectrum (dotted line) for *Ulva*. (From Blinks, *Annual Review of Plant Physiology*, Vol. 5, 1954)

of various wavelengths but of the same intensity). A typical result is shown in fig. 33 (dotted line).

If figs. 32 and 33 are studied together, several important points emerge:

    (i) The action spectrum closely follows the absorption spectrum for the entire chloroplast, suggesting that all the wavelengths are used in proportion to their absorption.

    (ii) The absorption by red and blue corresponds with absorption by the chlorophylls.

    (iii) The absorption in the mid-regions of the spectrum is mainly brought about by the carotenoids and this energy can also be utilized in photosynthesis.

It is a well-known fact that variegated leaves are not able to synthesize in the non-green parts, which often owe their pigment to the presence of carotenoids. The probable explanation of this is that the energy absorbed by the carotenoids can be transferred to the chlorophylls when these are *present in the same chloroplasts*, but that the carotenoids do not themselves have the ability to act directly in the photochemical reactions.

Further progress in understanding the nature of the light reaction stems from the work of the late Robert Emerson and his colleagues at the University of Illinois. They studied the absorption spectrum of

Chlorella in great detail and found that at wavelengths greater than 680 nanometres there is a very dramatic falling off in the efficiency of photosynthesis (they expressed their results in terms of 'quantum yield' which may be considered as the amount of photosynthesis taking place for each quantum of light absorbed) as can be seen in fig. 34.

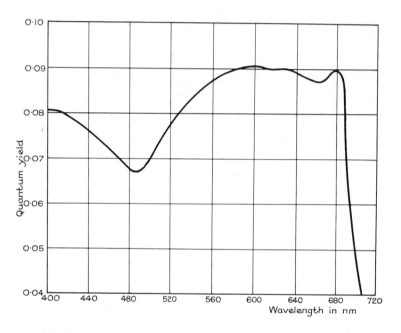

FIG. 34. The quantum yield of photosynthesis as a function of wavelength for *Chlorella* (after Emerson)

Since, at these wavelengths chlorophyll *a* is the only photosynthetic pigment absorbing light, it would seem that, for full photosynthesis, light of a shorter wavelength must also be absorbed, presumably by a different pigment. Emerson put this idea to the test by supplementing the 'far-red' light with light of a shorter wavelength (*c.* 650 nm).

If: Rate of photosynthesis when plant is illuminated at 680 nm = *a*
Rate of photosynthesis when plant is illuminated at 650 nm = *b*
Rate when illuminated at 680 nm *and* 650 nm *simultaneously* = *c*,
then Emerson found that *c* > (*a* + *b*).

The difference between the two sets of values [*c* − (*a* + *b*)] was called the '*enhancement effect*'. By maintaining a constant intensity and wavelength of far-red light and varying the shorter wavelength, he was able to plot the action spectrum of the enhancement effect.

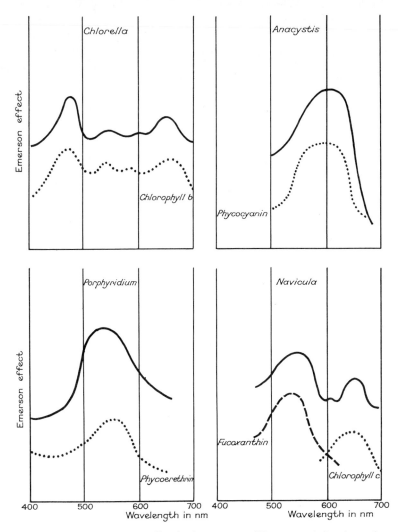

FIG. 35. The enhancement effect is shown as a solid curve and the absorption spectrum of the accessory pigment as a dotted line (modified after Emerson and Rabinowitch)

When this was done for a variety of different plants (fig. 35) it was found that there was a close correlation between the action spectrum and the absorption spectrum of the accessory pigments—chlorophyll

b for *Chlorella*, phycocyanin for *Anacystis* (Cyanophyceae), phycoerethrin for *Porphyridium* (Rhodophyceae) and fucoxanthin for *Navicula* (Bacillariophyceae).

The obvious explanation of this work is that photosynthesis requires *two* light-driven reactions, one operating with longer wavelength light and the other with light of shorter wavelengths.

The first working hypothesis was advanced by Hill and Bendall (1960) who suggested that electron transport in the chloroplast took place in the directions

$$H_2O \rightarrow \text{cytochrome } b_6 \rightarrow \text{cytochrome } f \rightarrow NADP$$

A study of the oxidation–reduction potentials involved shows that electrons travel 'uphill' to cytochrome b_6 (from $+0.8$ V $\rightarrow -0.06$ V), 'downhill' from cytochrome b_6 to cytochrome f (from -0.06 V $\rightarrow +0.365$ V) and then 'uphill' again to NADP (from $+0.365$ V $\rightarrow -0.32$ V). The two 'uphill' stages were thought to be energized by two separate light reactions and the 'downhill' stage could be coupled to the synthesis of ATP from ADP and inorganic phosphate.

There would now seem to be fairly general acceptance of the following:

(a) Light at 668 nm is optimally absorbed by a special form of chlorophyll (chlorophyll a-670) and this pigment may also receive energy from light absorbed by the accessory pigments.

(b) It is during this stage that oxygen is produced and an electron, displaced from the chlorophyll, combines with the start of the electron transport chain.

These two stages constitute Photosystem II.

(c) The components of the 'downhill' stage to cytochrome f are plastoquinone, cytochrome b_6 and plastocyanin (there is some controversy about the position of plastocyanin—it might be on the other side of cytochrome f.)

(d) In Photosystem I light is absorbed optimally at 683 nm by another special form of chlorophyll known as P_{700}.

(e) The immediate electron donor to P_{700} is either cytochrome f or plastocyanin. It was generally accepted that the electron energized by P_{700} was transferred to ferredoxin, but a certain amount of doubt was caused by the fact that the electrons appear to be over-energized—i.e. they could travel up a steeper gradient than that from P_{700}—to ferredoxin. It is now thought likely that the electrons are transferred to FRS (ferredoxin-reducing substance), with an oxidation–reduction potential of about -0.6 V before passing to ferredoxin (-0.42 V) and thence to NADP (-0.32 V). Thus during this stage the electrons are again travelling 'downhill' and ATP can be synthesized. The various stages involved are shown in Fig. 36.

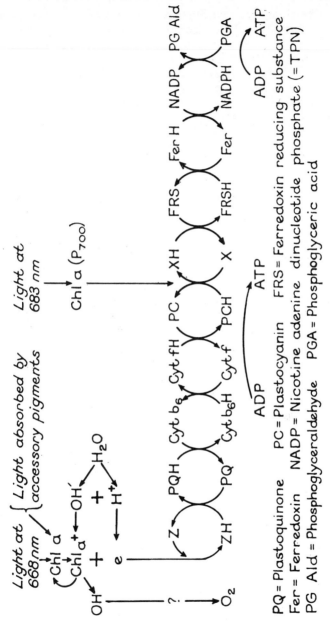

Fig 36. Summary of light reactions in photosynthesis.

PQ = Plastoquinone   PC = Plastocyanin   FRS = Ferredoxin reducing substance
Fer = Ferredoxin   NADP = Nicotine adenine dinucleotide phosphate (=TPN)
PG Ald = Phosphoglyceraldehyde   PGA = Phosphoglyceric acid

The Nature of the Dark (Light Independent) Reaction

In this stage the reducing power from water photolysis is utilized with the product of CO_2 fixation. The main method of investigation has been to illuminate cultures of acellular algae (e.g. *Chlorella*, *Scenedesmus*) in the presence of $^{14}CO_2$ for periods of time ranging from five seconds to several minutes before plunging them into boiling ethyl alcohol in order to kill them (and so stop any further reactions) and also to extract the product of CO_2 fixation. Obviously after only five seconds' exposure to $^{14}CO_2$ the radioactivity would be found in the products of the first reactions of the dark stages; killing after longer periods would show the fate of these early products.

Among the many compounds found are

Phosphoglyceric acid (a C_3 compound) after only 5 seconds' exposure

Phosphoglyceraldehyde ⎫
Dihydroxyacetone phosphate ⎬ after 30 seconds' exposure

Tetroses, pentoses, hexoses and heptoses after 90 seconds' exposure.

It was therefore established that the first products of the dark reaction were phosphoglyceric acid, phosphoglyceraldehyde and dihydroxyacetone phosphate. Since the conversion of acid to aldehyde is a reduction, it is presumably at this point that the reducing power from the light reaction is utilized.

If it is granted that the first product of photosynthesis is a C_3 compound, CO_2 fixation can be expressed as

$$CO_2 + X \longrightarrow C_3$$

Early attempts to identify X, the 'CO_2 acceptor', assumed a two-carbon compound, i.e.

$$CO_2 + C_2 \longrightarrow C_3$$

No appropriate compound was found and all the evidence points to the acceptor being a five carbon compound (ribulose diphosphate).

Among the evidence are experiments using $^{14}CO_2$ in which the relative concentrations of ribulose diphosphate and phosphoglyceric acid are measured when either the plant is placed in the dark and the level of CO_2 maintained constant or alternatively, the plant is kept in light of a constant intensity and the CO_2 removed.

The results of such experiments are shown in fig. 37 *a* and *b*. In particular it should be noted that:

(i) When placed in the dark the ribulose diphosphate level drops and that of the phosphoglyceric acid rises, pointing to the possibility that the phosphoglyceric acid is formed at the expense of the ribulose diphosphate.

(ii) When the CO_2 is removed there is a drop in the phosphoglyceric acid level and an increase in the amount of ribulose

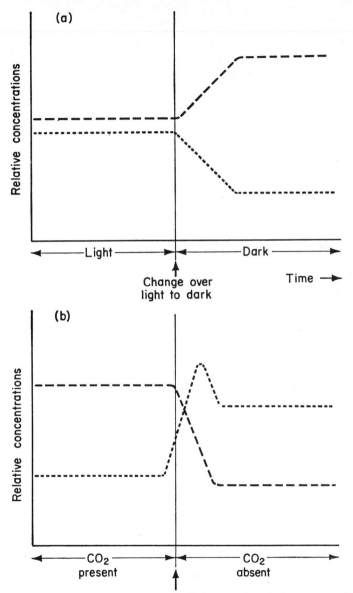

FIG. 37. Effect of (*a*) dark and (*b*) removal of CO_2 on the relative concentrations of
phosphoglyceric acid (– – –) and ribulose diphosphate (-------). (Redrawn
from Bassham, J. A., *J. Chem. Ed.*, **36**, No. 11, 548–554, 1959).

diphosphate suggesting that phosphoglyceric acid cannot be formed from ribulose diphosphate in the absence of CO_2.

(iii) (not shown on graph) There are no similar relationships when the concentrations of other substances are measured.

Thus it would seem reasonable to assume that CO_2 fixation involves the reaction

$$\begin{matrix} CH_2O\,\boxed{P} \\ | \\ CO \\ | \\ CHOH \\ | \\ CHOH \\ | \\ CH_2O\,\boxed{P} \end{matrix} \;+\; H_2O + CO_2 \;\longrightarrow\; 2\;\begin{matrix} CHO\,\boxed{P} \\ | \\ CHOH \\ | \\ COOH \end{matrix}$$

(Ribulose diphosphate)

From a commonsense point of view, as well as from experiments with $^{14}CO_2$, it would be expected that a regeneration of the C_5 compound would be involved in photosynthesis. The way in which this happens is similar to a well-known metabolic process—the pentose shunt—which plays a part in carbohydrate breakdown in plants and animals. The essentials of the scheme are shown in the carbon flow diagram of fig. 38.

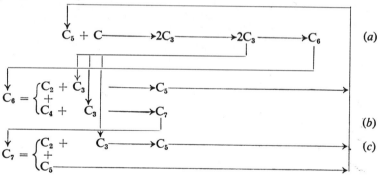

FIG. 38. Regeneration of the carbon acceptor. In line (a) the CO_2, probably as H_2CO_3, combines with the C_5 acceptor to form two molecules of phosphoglyceric acid which are reduced to a molecule of phosphoglyceraldehyde and a molecule of dihydroxyacetone phosphate (or rather an equilibrium mixture of the two). These are either able to form hexose (C_6) or are used in the regeneration process.

Line (b) summarizes the first stages of this regeneration. The two terminal C atoms of fructose phosphate are combined with phosphoglyceraldehyde to form xylulose phosphate (C_5) and erythrose phosphate (C_4). The erythrose phosphate reacts with dihydroxyacetone phosphate to form sedoheptulose phosphate (C_7).

The fate of the C_7 is shown in line (c). The two terminal C atoms react with phosphoglyceraldehyde to form another molecule of xylulose phosphate (C_5) and a molecule of ribose phosphate (C_5).

The three pentose molecules are then converted into ribulose diphosphate.

The overall reaction, in terms of carbon atoms, can be written cyclically as

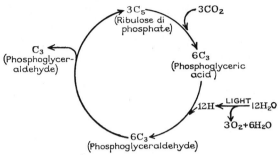

i.e. for every three molecules of acceptor used, three molecules are regenerated and one molecule of triose phosphate is available for the syntheses of carbohydrates, proteins, etc.

The fates of the phosphoglyceraldehyde and dihydroxyacetone phosphate are

(a) To combine together to form hexoses, from which polysaccharides (e.g. starch and cellulose) may be formed.

(b) To be converted, via pyruvic acid, into α ketoglutaric acid (in Krebs' cycle) from which glutamic acid and hence other amino acids can be produced.

(c) To form fats.

Outline reaction sequences are given in chapter 1 (pages 28–30) and the problems of nitrogen and fat metabolism are discussed in chapter 6.

The dark reactions outlined above do not operate in all plants and there has been a great deal of investigation carried out into CO_2 fixation by sugar cane, maize and some other graminae. It seems likely that CO_2 is first utilized by combining with phospho-enol pyruvic acid to produce oxaloacetic acid, and that this latter compound can easily be converted into malic or aspartic acid. Alternatively the COOH group on C_4 of the oxaloacetic acid can form the COOH group of phosphoglyceric acid which can be the starting point of sucrose synthesis.

Measurement of the Rate of Photosynthesis

This can be made by three methods:

(a) To measure the increase in dry weight.

(b) To measure CO_2 consumption.

(c) To measure O_2 evolution.

In all cases the measurement is related to changes/unit weight of tissue/unit time at a stipulated temperature. Allowances must be made for the effects of respiration, which will tend to decrease the

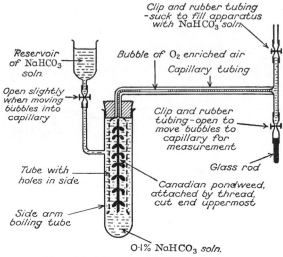

FIG. 39. Diagram of Audus's apparatus

values obtained, and this can be done by running a control experiment in the dark (assuming that light does not appreciably affect the rate of respiration).

Dry weight changes have the disadvantage of killing the tissue on which they are determined. The usual method is to cut out a circle of tissue from one side of a leaf midrib and determine its dry weight (by heating to constant weight at a temperature of about 110° C) and then comparing this value with that of a similar sized piece of tissue taken from a similar position on the other side of the midrib after a period of photosynthesis.

The measurement of CO_2 consumption is carried out by manometric methods as described on page 37.

The measurement of oxygen evolution is the simplest for class experiments. The most convenient apparatus is that designed by Audus in which the lengths of the bubbles* collected from photosynthesizing Canadian pondweed after a fixed time are measured in a capillary tube. The length of the bubble is found to be proportional to the rate of photosynthesis.

* The gas given off is *not* pure oxygen but oxygen-enriched air (about 25% O_2) from the intercellular spaces. The rate at which it is displaced is proportional to the rate of oxygen evolution.

The Influence of External Factors on the Rate of Photosynthesis. It should now be possible to deduce the effects of varying the external factors on the rate of photosynthesis if the following points are borne in mind:

(a) There is a light dependent stage in which

$$H_2O + NADP \longrightarrow NADP\ H_2 + \tfrac{1}{2}O_2$$

(b) There are light independent stages in which
    (i) Ribulose diphosphate $+ CO_2 \longrightarrow$ Phosphoglyceric acid.
    (ii) Phosphoglyceric acid $+ NADP\ H_2 \longrightarrow$ Triose phosphate $+ NADP$.
    (iii) Ribulose diphosphate is regenerated.
    (iv) Triose phosphate $\longrightarrow$ hexoses, polysaccharides, etc.

(c) The overall process involves an interaction of light and dark stages.

(d) Enzymes are thermolabile (see chapter 1, page 10) so that enzyme catalysed reactions
    (i) Are inhibited by temperatures greater than $c.$ 30° C due to the thermal inactivation of the enzymes.
    (ii) Have a $Q_{10} \simeq 2$, up to $c.$ 30° C.

(e) Enzymes act by combining with their substrate. When the substrate concentration is increased the graph of rate of reaction $v.$ substrate concentration takes the form of a right-angled hyperbola.

The external factors to be considered are

CO_2 concentration	Light intensity
Temperature	Water
Presence of various salts	O_2 concentration.

The effect of CO_2 concentration was first analysed extensively by Blackman and his school. At high light intensities and a constant temperature the rate of photosynthesis in water plants, e.g. *Elodea Canadensis*, was proportional to the concentration of CO_2 within a certain range, above which further increases in CO_2 concentration had no further effect. In the case of unicellular algae (e.g. *Chlorella*) under broadly similar conditions, Warburg obtained a curve which differed from that of Blackman in that it approximated closely to a rectangular hyperbola.

The differences are probably to be attributed to the proximity of the chloroplast to the measured external concentration of CO_2. In the case of *Chlorella* the distance is extremely small and the external CO_2 is virtually in direct contact with the chloroplast. As a result the rate of photosynthesis is governed by the combination of CO_2 with enzymes and so the graph is a right-angled hyperbola. In the case of

Elodea it is necessary for the CO_2 to reach the chloroplast from the external solution. At extremely high light intensities, this results in the rate of diffusion of CO_2 from the external solution being the limiting factor and it can be shown that this diffusion rate is linearly proportional to the concentration of CO_2—hence the straight slope in

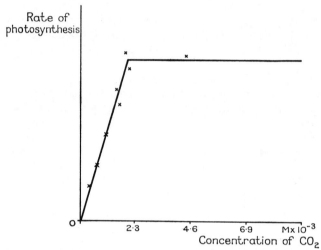

FIG. 40. Effect of CO_2 concentration on the rate of photosynthesis of *Elodea* at a constant high light intensity. (Based on the data of Blackman and Smith)

Blackman's graph. Eventually all the available light energy is utilized by the CO_2, so that after a certain concentration of CO_2, further increases in CO_2 concentration are unable to have any effect on the rate of photosynthesis for a given light intensity.

The interdependence of light intensity and CO_2 concentration is further shown in the results of Harder for *Fontinalis*. The form of these graphs is intermediate between the extremes of Warburg and Blackman. The essential feature to notice is that for any given light intensity a point is reached where increasing the CO_2 concentration does not affect the rate of photosynthesis, but if a higher light intensity is used, then a higher CO_2 concentration can be utilized.

The effect of light intensity. The intensity of light affects the rate of the photolytic reaction and so controls the rate of production of ATP and reduced coenzyme. Hence for a given concentration of CO_2, varying the light intensity produces a rectangular hyperbola graph similar to that already discussed for the effect of CO_2. This is best shown in measurements with unicellular green algae. In the case of higher plants other factors may also be involved among which may be mentioned the movement of the chloroplasts in response to increased light intensity and selective absorption of light by the cell layers of the leaf.

The effect of temperature is on the velocity of enzyme-controlled reactions in the dark stage. When the light intensity is low, the reaction is limited by the small quantities of reduced coenzymes available so that temperature increases have little effect on the overall rate.

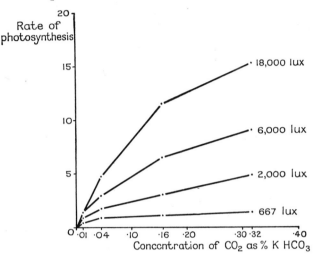

FIG. 41. Effect on photosynthesis of various CO_2 concentrations at different light intensities. (Redrawn from data of Harder)

At high light intensities, on the other hand, it is the enzyme-controlled dark stage which controls the rate of photosynthesis and there is a $Q_{10} \simeq 2$. If the temperature is greater than about 30° C, thermal inactivation of enzymes results so that the overall increase caused by higher temperatures is more than offset by the reduction in the quantities of available active enzymes. This is strikingly shown in the results of Matthaei for Cherry Laurel.

The continuous line shows an apparent optimum at *c.* 37° C, when measurements are made after two hours at that temperature. If, however, the plant is maintained for five hours at a given temperature, the increased photosynthetic rate is only maintained at temperatures below 25° C and the optimum would now appear to be at 30° C (broken line). Obviously these latter temperatures give a more accurate picture of conditions which a plant would encounter in the field, and so it would appear that the true optimum temperature is less than that sometimes suggested.

In the case of low light intensities, the equivalent graph would run parallel to the abscissae until about 30° C, after which it would gradually fall as the inactivated enzymes bring the dark stage to a halt.

The effect of water would rarely seem to be a controlling factor as normally the chloroplasts appear to contain an abundant supply of water. Many observations suggest that in the field the plant is able to

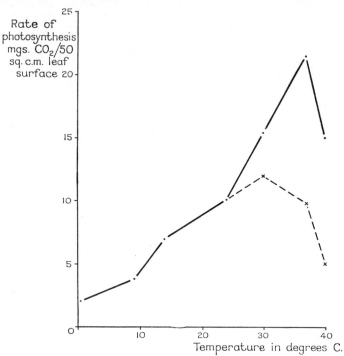

Fig. 42. Variation in rate of photosynthesis for Cherry Laurel with temperature. Solid line: measurements after 2 hours at that temperature. Dotted line: after 5 hours. Up to 25° C, rate was steady for the 5 hours. (Based on data of Matthaei)

tolerate a wide range of soil moisture without any significant effects on photosynthesis, and it is only when wilting sets in that photosynthesis is retarded. This is probably not due to actual lack of water in the chloroplasts but rather because of loss of turgor in the guard cells and consequent stomatal closing (see chapter 1, page 39).

The effect of mineral salts can only be mentioned here in general terms as the literature on this aspect is very complex. The most outstanding effects are caused by deficiencies in Fe, Mn and Mg which result in a low chlorophyll level (chlorosis) and so limit the light reaction. Other elements which also cause effects via the dark stage include phosphorus, presumably by limiting the production of ATP, and sulphur, affecting enzyme production.

A more detailed account of the functions of mineral salts is given in chapter 7.

The effect of O_2 concentration

Warburg found that if the concentration of O_2 was raised above that of air, there was a reduction in the rate of photosynthesis. More recently Björkman (1967) has shown that the converse relationship also holds—viz. reducing the oxygen concentration results in increased rates of photosynthesis. The mechanism is unknown but it is possible that oxygen may act as an inhibitor by combining with some of the intermediates of the electron—hydrogen transfer chains.

6

Nitrogen and Fat Metabolism

For the sake of convenience the metabolism of nitrogenous compounds and of fats are considered separately from photosynthesis, but it must be emphasized again that the formation of amino acids and fats really form an integral part of photosynthesis.

NITROGEN METABOLISM

This topic falls into two main subdivisions—the formation of amino acids and the subsequent synthesis of these into proteins. Many aspects of amino acid synthesis are peculiar to green plants but their elaboration into proteins raises problems which are fundamental to the physiology of all living organisms.

Amino Acid Synthesis

In the account of photosynthesis given in the preceding chapter, emphasis was placed on the fact that phosphoglyceraldehyde can be considered as the primary product. This phosphorylated triose differs from even the simplest amino acid in the lack of any nitrogen atoms in the molecule: the means of incorporation of these will now be described.

Sources of Nitrogen. The most abundant source is gaseous nitrogen in the atmosphere. Also important are nitrate ions (NO_3'), ammonium (NH_4^+), undissociated NH_3 in solution and, in the case of insectivorous plants, insect protein. There is also an increasing body of evidence to suggest that some plants are able to utilize directly the nitrogen found in urea ($CO(NH_2)_2$).

Of these various sources the most important to the majority of plants is the nitrate ions in the soil solution.

The Utilization of Nitrate Ions. Nitrate ions are absorbed across the root hairs by the process of anion respiration and are accumulated as free ions in the cell vacuoles. Before they can be utilized it is almost certain that they are reduced to ammonia, probably by the pathway*

Nitrate $\rightarrow$ Nitrite $\rightarrow$ Hydroxylamine $\rightarrow$ Ammonia

* There is evidence (Fewson and Nicholas, *Nature*, **188** (1960), and *Nature*, **190** (1961)), to suggest that nitric oxide and nitroxyl may be involved as intermediates, i.e.

$HNO_3 \rightarrow HNO_2 \rightarrow NO \rightarrow (NOH) \rightarrow NH_2OH \rightarrow NH_3$

In favour of such a pathway may be mentioned

(i) The possible substitution of ammonium ions for nitrate without in any way affecting amino acid synthesis (providing that care is taken to regulate the pH of the external solution).

(ii) Eckerson's analysis of tomato plants with an initially low nitrogenous content and an absence of nitrate. When such plants were supplied with nitrate there was first an accumulation of nitrate throughout the plant which gradually decreased in amount as there was a progressive increase in first nitrites, then ammonia and finally amino acids. The results are summarized below

Distribution of	After 24 hr	After 36 hr	After 48 hr	After 5–10 days
Nitrate	Throughout plant	Decrease	Decrease	Trace
Nitrite	Absent	Throughout plant	Decrease	Trace
Ammonia	Absent	Trace	Throughout plant	Trace
Amino acids	Not detached	Not detached	Not detached	Throughout plant

(iii) The existence of the appropriate reductase enzymes in flowering plants. The reactions involved are

$$NO_3' + 2[H] \xrightarrow{\text{Nitrate reductase}} NO_2' + H_2O$$

$$NO_2' + 5[H] \xrightarrow{\text{Nitrite reductase}} NH_2OH + H_2O$$

$$NH_2OH + 2[H] \xrightarrow[\text{reductase}]{\text{Hydroxylamine}} NH_3 + H_2O$$

The hydrogen *donor* is NADP H_2 (or, in the case of nitrate and hydroxylamine reductases, NAD H_2 may provide an alternative). The hydrogen *source* may be linked with either photolysis or respiration. The latter is the only possible means when the reductions take place in non-photosynthetic tissue (e.g. the root) but photolysis may be involved when reductions take place in green parts of the plant.

The incorporation of ammonia takes place by a process of 'reductive amination'. In the flowering plants this is limited to the reductive amination of α ketoglutaric acid. This is one of the Krebs' cycle acids and can be produced from triose phosphate or other carbohydrates by the pathway outlined on page 29. It reacts with ammonia to form glutamic acid.

$$
\begin{array}{c}
\text{COOH} \\
| \\
(\text{CH}_2)_2 \\
| \\
\text{CO} \\
| \\
\text{COOH}
\end{array}
+ \text{NH}_3 \quad \rightleftharpoons \quad
\begin{array}{c}
\text{COOH} \\
| \\
(\text{CH}_2)_2 \\
| \\
\text{CNH} \\
| \\
\text{COOH}
\end{array}
+ \text{H}_2\text{O}
$$

α Ketoglutaric α Iminoglutaric
acid   acid

$$
\begin{array}{c}
\text{COOH} \\
| \\
(\text{CH}_2)_2 \\
| \\
\text{CNH} \\
| \\
\text{COOH}
\end{array}
+ 2[\text{H}] \quad \rightleftharpoons \quad
\begin{array}{c}
\text{COOH} \\
| \\
(\text{CH}_2)_2 \\
| \\
\text{CHNH}_2 \\
| \\
\text{COOH}
\end{array}
$$

Glutamic
acid

The Formation of other Amino Acids. Glutamic acid, since it is the only one formed by reductive amination from ammonia and an α keto acid, is referred to as a *primary* amino acid. Other amino acids are formed from glutamic acid by various processes, among which *transamination* is particularly important. Amino acids formed by transamination from glutamic acid are referred to as *secondary* amino acids.

As examples of transamination (see also page 18) we may consider the reactions between glutamic acid and oxaloacetic acid and also between glutamic acid and pyruvic acid.

$$
\begin{array}{c}
\text{COOH} \\
| \\
(\text{CH}_2)_2 \\
| \\
\text{CHNH}_2 \\
| \\
\text{COOH}
\end{array}
+
\begin{array}{c}
\text{COOH} \\
| \\
\text{CH}_2 \\
| \\
\text{CO} \\
| \\
\text{COOH}
\end{array}
\xrightleftharpoons[\text{transaminase}]{\substack{\text{Glutamic acid}/ \\ \text{oxaloacetic acid}}}
\begin{array}{c}
\text{COOH} \\
| \\
(\text{CH}_2)_2 \\
| \\
\text{CO} \\
| \\
\text{COOH}
\end{array}
+
\begin{array}{c}
\text{COOH} \\
| \\
\text{CH}_2 \\
| \\
\text{CHNH}_2 \\
| \\
\text{COOH}
\end{array}
$$

Glutamic   Oxaloacetic α Ketoglutaric   Aspartic
acid   acid   acid   acid

$$
\begin{array}{c}
\text{COOH} \\
| \\
(\text{CH}_2)_2 \\
| \\
\text{CHNH}_2 \\
| \\
\text{COOH}
\end{array}
+
\begin{array}{c}
\text{CH}_3 \\
| \\
\text{CO} \\
| \\
\text{COOH}
\end{array}
\xrightleftharpoons[\text{transaminase}]{\substack{\text{Glutamic acid}/ \\ \text{pyruvic acid}}}
\begin{array}{c}
\text{COOH} \\
| \\
(\text{CH}_2)_2 \\
| \\
\text{CO} \\
| \\
\text{COOH}
\end{array}
+
\begin{array}{c}
\text{CH}_3 \\
| \\
\text{CHNH}_2 \\
| \\
\text{COOH}
\end{array}
$$

Pyruvic   Alanine
acid

In experiments in which the plant is fed with heavy nitrogen (as

$^{15}NO_3{'}$ or $^{15}NH_4{}^+$) it is found that the ^{15}N is incorporated first into glutamic acid and, very shortly afterwards, into aspartic acid and alanine—in fact the difference in time is so small that it has only recently been found possible to observe it experimentally. Aspartic acid and alanine are secondary amino acids.

Extensive chemical analyses have revealed the presence, in plant tissues, of a large number of α keto acids which correspond to all the naturally occurring α amino acids. These α keto acids are manufactured by a variety of complex reactions—glucose oxidations, organic acid metabolism and intermediates in aromatic biosyntheses. They can undergo transaminations with either glutamic acid or the secondary amino acids to form their homologous amino compounds.

The whole process is essentially cyclical, since only the α ketoglutaric acid can combine with ammonia.

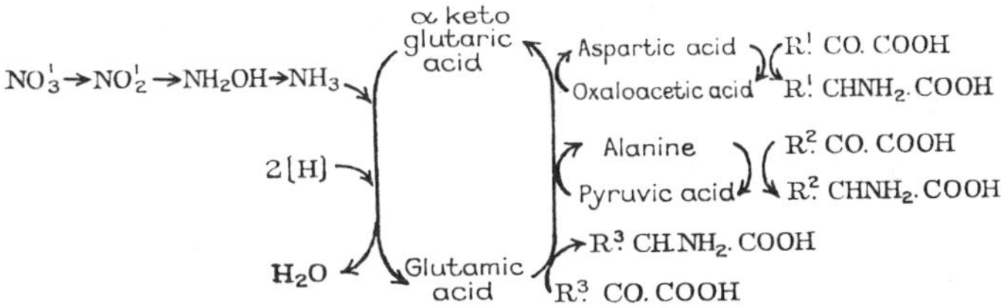

FIG. 43. Role of reductive amination and transaminations in amino acid synthesis. (R′CO·COOH, R²CO·COOH, R³CO·COOH represent α keto acids which form, by transamination, homologous α amino acids)

The Role of Amides—Glutamine and Asparagine. The amides glutamine and asparagine are frequently found in plant tissues and are formed by the reaction of ammonia with the appropriate amino acid, e.g.

$$
\begin{array}{ccc}
\text{COOH} & & \text{CONH}_2 \\
| & & | \\
(\text{CH}_2)_2 & & (\text{CH}_2)_2 \\
| & +\ \text{NH}_3 \ \rightleftharpoons & | \qquad +\ \text{H}_2\text{O} \\
\text{CHNH}_2 & & \text{CHNH}_2 \\
| & & | \\
\text{COOH} & & \text{COOH} \\
\text{Glutamic} & & \text{Glutamine} \\
\text{acid} & &
\end{array}
$$

$$
\begin{array}{ccc}
\text{COOH} & & \text{CONH}_2 \\
| & & | \\
\text{CH}_2 & & \text{CH}_2 \\
| & +\ \text{NH}_3 \ \rightleftharpoons & | \qquad +\ \text{H}_2\text{O} \\
\text{CHNH}_2 & & \text{CHNH}_2 \\
| & & | \\
\text{COOH} & & \text{COOH} \\
\text{Aspartic} & & \text{Asparagine} \\
\text{acid} & &
\end{array}
$$

An important function of these amides is the provision of a means of storing excess nitrogen. This can be utilized by their reaction with α keto acids.

$$
\begin{array}{c}
\text{CONH}_2 \\
| \\
(\text{CH}_2)_2 \\
| \\
\text{CHNH}_2 \\
| \\
\text{COOH} \\
\text{Glutamine}
\end{array}
\;+\;
\begin{array}{c}
\text{R}' \\
| \\
\text{CO} \\
| \\
\text{COOH} \\
\alpha\,\text{Keto} \\
\text{acid}
\end{array}
\;+\;2[\text{H}]\;\rightleftharpoons\;
\begin{array}{c}
\text{COOH} \\
| \\
(\text{CH}_2)_2 \\
| \\
\text{CHNH}_2 \\
| \\
\text{COOH} \\
\text{Glutamic} \\
\text{acid}
\end{array}
\;+\;
\begin{array}{c}
\text{R}' \\
| \\
\text{CHNH}_2 \\
| \\
\text{COOH} \\
\text{Homologous} \\
\alpha\,\text{amino acid}
\end{array}
$$

Their formation and functioning as a nitrogen store is shown in fig. 44.

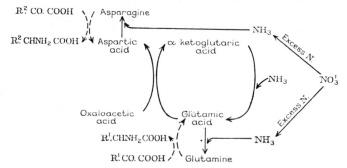

F I G. 44. The formation of amides (continuous lines) and their functioning as a means of supplying NH₂ groups (broken lines)

In addition, the amides may also act as a means of translocating NH_2 groups during the mobilization of storage protein for growth. The protein is first hydrolysed, under the influence of proteases, into its constituent amino acids which are then deaminated by the transference of their amino groups first to α ketoglutaric acid (or oxaloacetic acid) and then to glutamic (or aspartic) acid, e.g.

Protein + H₂O $\xrightarrow{\text{Proteases}}$ Amino acids

→ Amino acid + α ketoglutaric acid → α keto acid + glutamic acid

→ Amino acid + glutamic acid   → α keto acid + glutamine

The α keto acids are utilized as respiratory substrates and the amide is translocated to the growing points where, with newly synthesized α keto acids (see above) the reverse series of reactions can take place. This is summarized in fig. 45.

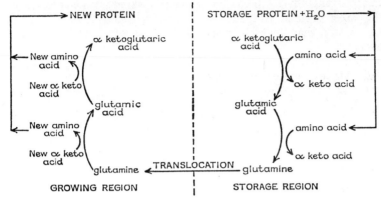

FIG. 45. Function of amides in translocation

In the case of etiolated tissues a similar process occurs (see fig. 46). Amino acids resulting from protein hydrolysis are deaminated by transamination with either oxaloacetic or α ketoglutaric acids. The resulting aspartic or glutamic acid is then converted into the corresponding amide at the expense of ammonia from more amino acids. The keto acids formed are utilized as a respiratory substrate.

On returning the plant to the light, α keto acids are synthesized by

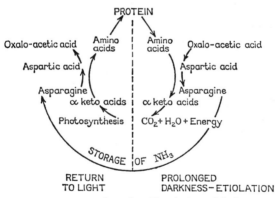

FIG. 46. Function of amides during etiolation

photosynthesis and these are converted into amino acids by the reconversion of asparagine (or glutamine) to oxaloacetic (or α ketoglutaric) acid.

Other Sources of Nitrogen. Insectivorous plants appear to be able to utilize amino acids directly by absorption through their leaves. Little recent work has been recorded on this aspect of nitrogen supply but it is generally accepted that the mechanism is first a hydrolysis of the insect protein, catalysed by proteases secreted by glands in the leaves, and that this is followed by foliar absorption of the amino acids.

Urea also enters the plant by foliar absorption and this has considerable importance in agriculture.

Many plants possess the enzyme urease which catalyses the reaction

$$CO(NH_2)_2 + H_2O \rightleftharpoons CO_2 + 2NH_3$$

At first sight it would appear likely that urea would function as an alternative source of ammonia. That this is not necessarily the case is strongly suggested by two pieces of evidence:

(a) When the C of the administered urea is isotopically labelled —i.e. $^{14}CO(NH_2)_2$, the ^{14}C is incorporated, together with the N, especially into glutamine.

(b) In plants lacking the enzyme urease, urea absorption can still occur.

In mammalian liver, the existence of an 'ornithine cycle' has been conclusively demonstrated as a method of converting ammonia, resulting from amino acid deamination, into urea. The essential stages of this cycle are

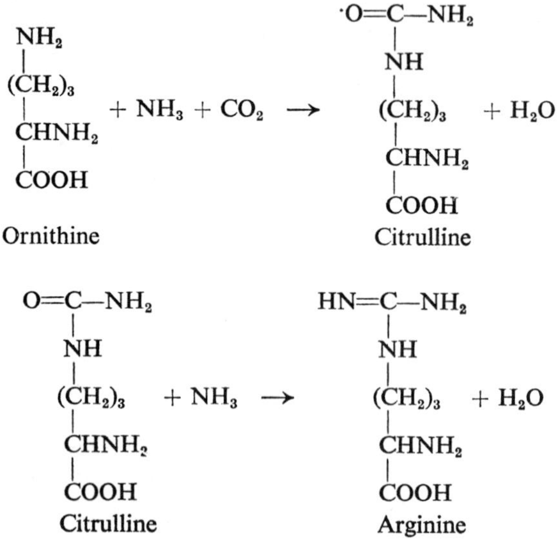

$$
\begin{array}{l}
NH_2 \\
| \\
(CH_2)_3 \\
| \\
CHNH_2 \\
| \\
COOH
\end{array}
+ NH_3 + CO_2 \rightarrow
\begin{array}{l}
\cdot O{=}C{-}NH_2 \\
| \\
NH \\
| \\
(CH_2)_3 \\
| \\
CHNH_2 \\
| \\
COOH
\end{array}
+ H_2O
$$

Ornithine   Citrulline

$$
\begin{array}{l}
O{=}C{-}NH_2 \\
| \\
NH \\
| \\
(CH_2)_3 \\
| \\
CHNH_2 \\
| \\
COOH
\end{array}
+ NH_3 \rightarrow
\begin{array}{l}
HN{=}C{-}NH_2 \\
| \\
NH \\
| \\
(CH_2)_3 \\
| \\
CHNH_2 \\
| \\
COOH
\end{array}
+ H_2O
$$

Citrulline   Arginine

$$HN=C-NH_2$$
$$|$$
$$N-H$$
$$|$$
$$(CH_2)_3 \quad + H_2O \quad \rightarrow$$
$$|$$
$$CHNH_2$$
$$|$$
$$COOH$$

Arginine

$$O=C\begin{array}{c}NH_2\\NH_2\end{array} \quad \text{Urea}$$
$$+$$
$$NH_2$$
$$|$$
$$(CH_2)_3$$
$$|$$
$$CHNH_2$$
$$|$$
$$COOH$$

Ornithine

This is shown simply in fig. 47.

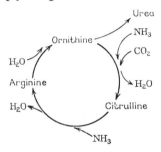

FIG. 47.

The major stages have recently been demonstrated in plants, although the cyclical operation has not yet been conclusively shown. If it does occur, then a possible method of utilizing urea exists as suggested by Walker for *Chlorella*, viz. a condensation of urea with ornithine to form arginine, i.e. the cycle working in reverse.

$$O=C\begin{array}{c}NH_2\\NH_2\end{array} \quad + \begin{array}{c}NH_2\\|\\(CH_2)_3\\|\\CHNH_2\\|\\COOH\end{array} \quad \rightarrow \quad \begin{array}{c}HN=C-NH_2\\|\\N-H\\|\\(CH_2)_3\\|\\CHNH_2\\|\\COOH\end{array} \quad + H_2O$$

Ornithine   Arginine

It has been shown for mammalian tissue that in the conversion of citrulline to arginine, arginosuccinic acid is an intermediate formed by the combination of citrulline with aspartic acid. The reactions are:

Citrulline + Aspartic acid $\rightarrow$ Arginosuccinic acid
Arginosuccinic acid $\rightarrow$ Arginine + Fumaric acid

If this reaction sequence takes place in reverse, then it becomes possible for the nitrogen of urea to be incorporated into aspartic acid.

Urea + Ornithine $\longrightarrow$ Arginine $\longrightarrow$ Arginosuccinic acid $\longrightarrow$ Citrulline

$\nearrow$ Fumaric acid

$\searrow$ Aspartic acid

The aspartic acid could then be converted into glutamic acid, by a transamination with α ketoglutaric acid, and thence into glutamine.

Atmospheric Nitrogen. In addition to the supply of inorganic nitrogen already referred to, an abundant supply exists in the form of molecular nitrogen in the earth's atmosphere. The ability to utilize this source directly is found in some bacteria, blue-green algae (Cyanophyceae) and fungi. As an example of nitrogen fixation only that occurring in bacteria will be considered here, particularly those in symbiotic association with higher plants.

The ability of leguminous plants to revitalize the soil after a cereal crop has been the basis of crop rotation in agricultural practice for several centuries. Modern work on this subject stems from the experiments of Boussingault in 1838 when he showed that if clover plants were grown in sand their nitrogen content increased but that of wheat, under similar conditions, did not.

Subsequently, in 1888, Hellriegal and Wilfarth related the occurrence of nodules—resulting from bacterial invasion—on the roots with the ability to utilize atmospheric nitrogen. They found that as long as nodules were present legumes were able to increase their nitrogen content even in the absence of any supply of 'fixed' nitrogen, e.g. nitrates or ammonium salts. If, on the other hand, plants were grown in completely sterile conditions, nodules did not develop and growth was restricted without an external supply of nitrate or ammonium. They also introduced the conception that the relationship between the nodule bacteria (at that time classed as *Bacterium radicicola*) and the host plant was essentially symbiotic, the bacteria providing the legume with a nitrogenous supply.

B. radicicola has now been renamed and placed in the genus *Rhizobium*. This genus is divided into species on the basis of the species of legume which can be inoculated, since any one type of *Rhizobium* is restricted in the number of legume species which it can infect.

When the *Rhizobium* is free in the soil it has a coccoid form which can develop into swarmers swimming in the soil moisture. The swarmers secrete auxins, causing root hairs to curl and become distorted and an infection thread is produced which enters the hair and penetrates into the deeper layers of the cortex. The invaded cells and their neighbours are stimulated to divide and as a result the characteristic nodules are formed whilst the bacteria become swollen and branched: in this stage they are known as bacteroids.

If detached nodules are exposed to $^{15}N_2$ for varying lengths of time and then fractionated in the cold it is found that most of the

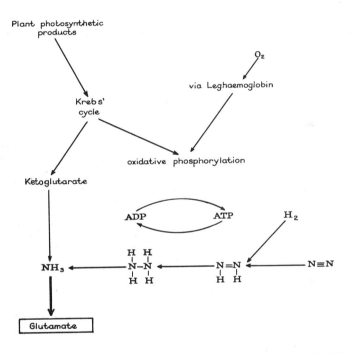

FIG. 48. An outline scheme of legume-nodule nitrogen fixation (simplified from F. J. Bergersen, 1971)

label accumulates in the soluble fraction—during the first minute as NH_3 but shortly afterwards in the α amino groups of glutamic acid and glutamine—and much less, if any, labelling is found in the bacteroids. Anaerobically prepared and washed bacteroid suspensions are also capable of fixing N_2 into NH_3 when supplied with lower partial pressures of oxygen so it appears likely that the plant component is not directly involved in this part of the reaction. By the disruption of dense suspensions it has been found possible to isolate an enzyme—nitrogenase—and comparisons of nitrogenase activities from different sources suggest that they are all essentially similar. They are all capable of catalysing the reaction $N_2 \rightarrow NH_3$ and they all have an absolute requirement for ATP.

A feature common to many nodules is the presence of a form of haemoglobin known as Leghaemoglobin. It has an extremely high affinity for oxygen and it is thought to play an indirect part in nitrogen fixation by accelerating the diffusion of oxygen into the nodules so that the oxygen demand of ATP synthesis is met.

It is not possible to give a definitive account of all the pathways involved but fig. 48 does give some possible outlines.

Before discussing the actual synthesis of protein from the amino acids—a topic which, as has already been mentioned, is of paramount importance to both plant and animal physiology—it is pertinent to consider the overall balance of nitrogen in nature.

The Nitrogen Cycle. A picture has developed in which the nitrates and ammonium compounds of the soil are absorbed by the plants and, together with the products of photosynthesis, converted into amino acids and proteins, so that the inorganic nitrogenous salts are 'locked' in the plant protoplasm. In passing it should be noted that the operation of animal food chains results in some of this nitrogen entering into animal protoplasm.

In order to maintain a balance, it is necessary for this nitrogen to be unlocked when the plant or animal dies and for it eventually to be reconverted back to nitrates.

The first step involves the liberation of ammonia from the amino acids. The enzymes involved are amino acid oxidases and the reaction may be written as

$$\begin{array}{ccc}
R & & R \\
| & & | \\
CHNH_2 + [O] & \rightarrow & CO \quad + NH_3 \\
| & & | \\
COOH & & COOH
\end{array}$$

The enzymes can come from several sources.

(a) The plant or animal (i.e. autolysis or self-digestion).
(b) Soil fungi, e.g. *Penicillium* spp., *Aspergillus* spp., *Neurospora* spp.
(c) Soil bacteria, e.g. *Pseudomonas* spp.

The NH_3 is then converted first into nitrites and then into nitrates by the action of soil bacteria, *Nitrosomonas* and *Nitrobacter* respectively, which obtain their energy supply by carrying out the appropriate oxidations, viz.

Nitrosomonas

$$NH_3 + 3[O] \rightarrow HNO_2 + H_2O + 79 \text{ kcal}$$

Nitrobacter

$$HNO_2 + [O] \rightarrow HNO_3 + 21 \cdot 6 \text{ kcal}$$

In this way a supply of ammonium salts and nitrates is returned to the soil. The process is complicated by the presence of denitrifying bacteria which reduce the nitrates to gaseous nitrogen. Various pathways are possible: as an example *Pseudomonas denitrificans* acts via

$$HNO_3 \rightarrow HNO_2 \rightarrow HNO \rightarrow N_2$$

Nitroxyl

$$H_2N_2O_2$$
Hyponitrous acid

As a result of denitrification a loss of 'fixed' nitrogen occurs from the soil. The methods by which this loss can be made good have been largely considered, viz. the nitrogen-fixing ability of soil bacteria, saprophytic bacteria, soil fungi and blue-green algae. In addition may be mentioned the effect of thunderstorms where, at the temperature of the electric arc (lightning) nitrogen and oxygen combine to form nitric oxide which combines with more oxygen and water to produce nitric acid.

The complete nitrogen cycle is shown diagrammatically in fig. 49.

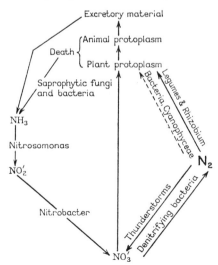

FIG. 49. The nitrogen cycle

Protein Synthesis. It is now well established that the proteins of any species, probably of any individual, are unique to that species or individual—i.e. they are highly specific. This specificity is thought to be a consequence of the sequence in which the amino acids are present in the proteins. Therefore any theory of protein synthesis must include an explanation of a mechanism by which the amino acids come to occupy a definite sequence in the protein. It will be recalled that enzymes are also proteins and so their synthesis must also be subjected to such a mechanism.

One of the earliest theories of protein synthesis was that it took place by a reverse of protein hydrolysis. The sequence would be:

$$\text{Amino acids} \longrightarrow \text{Polypeptides} \longrightarrow \text{Protein}$$

The evidence in favour of such a reaction is

(a) A tendency for protein synthesis to be associated with desiccation. In experiments with wheat grains, Eckerson found that there was no protein synthesis in the endosperm of full grown, still green, wheat kernels containing 90% moisture, but that after drying, there was an increase in the protein content with a gradual decrease in the amide content.

(b) The normal reversibility of enzyme catalysed reactions. Since the hydrolysis is exothermic, the equilibrium lies far towards amino acid production, and in experiments to artificially synthesize proteins by such a mechanism, the resultant products have not been characteristic of normal protein.

The main stumbling block to this theory is the large number of different types of proteins. Presumably each of these would require a specific enzyme and since, as mentioned above, the enzymes are also proteins, it is necessary to postulate a further set of highly specific enzymes to control the synthesis of the first set—either there is an unlimited number of enzymes, which seems highly unlikely, or the theory is untenable.

Most modern theories invoke the action of Ribose Nucleic Acid (RNA) as a template which is able to provide a means of incorporating the amino acids in a definite sequence into the protein. The idea of an association between RNA and protein synthesis comes from several lines of enquiry, dealing with both plant and animal tissues, among which may be mentioned

(a) A general correlation between the protein synthesizing activity of a cell and its RNA content. Cells in which there is a rapid protein production have a high concentration of RNA.

(b) The effect of the enzyme ribonuclease which digests RNA. Addition of ribonuclease to actively protein synthesizing tissue completely blocks protein formation.

(c) In the case of onion root tips, Brachet has further shown that the addition of RNA to cells, which have already had their protein synthesis inhibited by ribonuclease, results in a renewal of protein synthesis. A suggestion of specificity comes from the demonstration that the effect of RNA extracted from onion tissues is approximately three times as effective as yeast RNA.

(d) The high concentration of RNA in the ribosomes, which are also characterized by high concentrations of protein.

If it is accepted that RNA controls the synthesis of protein in the ribosomes of the cytoplasm, two further problems must also be answered—how is this synthesis controlled by the nucleus (studies in the no-man's-land between genetics, cytology, physiology and biochemistry all point to the concept that the inherited factors or genes largely operate by directing the synthesis of specific enzymes which operate outside the nucleus), and how is it that there is sufficient variation in the relatively simple composition of the nucleic acids to allow for the many templates required?

The nucleic acid which forms an intrinsic part of chromosome structure is desoxyribose nucleic acid (DNA). Its structure can be shown to consist of two long polymers each consisting of a backbone of desoxyribose molecules joined by phosphate. Attached to each desoxyribose is a base—either a purine (adenine or guanine) or a pyrimidine (cytosine or thymine). The sizes of the various components indicate that the bases of each polymer can only pair in one way—viz. adenine with thymine and cytosine with guanine. These groups are joined together by hydrogen bonds and the two polymers are twisted around each other to give a geometrical pattern which can best be described as a ladder twisted into a spiral. Fig. 50 shows a simplified diagram of the pattern of the various components but ignoring the spiral.

The generally held view is that it is the sequence of bases on the DNA molecule which determines the amino acid sequence for incorporation into a protein molecule and that the code involves three successive bases (triplets) which is read from one end of the molecule, e.g.

$$\underbrace{\text{A C T}}_{1} \quad \underbrace{\text{G A C}}_{2} \quad \underbrace{\text{G T T}}_{3} \quad \underbrace{\text{C G A}}_{4}$$

Since we have four cyphers and we only interpret in sequences of three, it follows that 64 possible pieces of information could be transmitted but there are only about twenty amino acids to be coded. It is now known that *all* the combinations have a meaning (i.e. there are no 'nonsense triplets'). The first two bases of a triplet are particularly important in coding for an amino acid (page 117). On this theory the classical gene would correspond to sequences of up to several thousand bases.

It is still necessary to account for this information being supplied to the ribosomes. RNA differs from DNA in several respects.

(*a*) It contains the pentose sugar *ribose* instead of desoxyribose.

(*b*) It contains the pyrimidine base *uracil* instead of thymine.

(*c*) It is essentially a single-stranded polymer (although it may be twisted on itself, like a hairpin, and then spiralled, as in transfer RNA).

(*d*) Three distinct forms exist—*messenger* RNA, *transfer* RNA and *ribosomal* RNA.

The transfer of information from chromosomal DNA into actual protein synthesis takes place in three relatively distinct phases.

(i) *Messenger RNA* becomes coded in the nucleus, i.e. the base

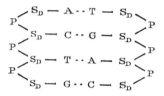

FIG. 50. Base pairs of DNA.

(S_D = desoxyribose, P = phosphate, A = adenine, G = guanine, C = cytosine
T = thymine)

sequence of the DNA becomes incorporated into a similar code on the RNA. There is some doubt as to whether this code transference is of the same type as that involved in the replication of DNA but with the substitution of uracil for thymine, viz.

DNA base sequence A—C—T—G
∴ RNA ,,   ,,   U—G—A—C

Alternatively it may happen that there is one RNA base coded for each *pair* of the DNA bases: if this is the case it might be expected that adjacent base pairs are also involved as a means of determining the 'sense' of a base pair—viz. to distinguish between A : T and T : A.

(ii) Whatever the details of the code transference are, the messenger RNA passes to the ribosomes. Ribosomal RNA, possibly originating from the nucleolus, is involved in the 'reading' of the code sequence of the messenger RNA.

(iii) About twenty different types of transfer RNA exist in the cytoplasm and each is specific for *one* amino acid. In addition an equivalent number of 'activating' enzymes exist which are able to catalyse, in the presence of ATP, the formation of an amino acid–RNA complex. If AA_1 represents an amino acid, E the appropriate transfer enzyme and $tRNA_1$ the transfer RNA for that particular amino acid, then the reaction sequence can be represented as

$$AA_1 + E_1 + ATP \rightarrow \{E_1-AMP-AA_1\} + 2P \curvearrowright 2\sim \textcircled{P}$$

$$tRNA_1 + E_1-AMP-AA_1 \rightarrow \{tRNA_1-AA_1\} + AMP + E_1$$

The AA–RNA complex has a high energy level. It migrates to the ribosomes and the transfer RNA is only able to come into contact with the messenger RNA at a point dictated by the coding on their two surfaces. When it does so, it leaves the amino acid in a position appropriate to the sequence it will occupy in the new protein molecule.

Confirmation of the broad principles described above comes from experiments with synthetic RNA. In the first experiments molecules of RNA were made in which uracil was the only base present so that the base sequence was –U–U–U. When this RNA was introduced into a mixture of amino acids it was found that a polypeptide was formed containing only one type of amino acid—phenylalanine. Further experiments along these lines involved the synthesis of RNA molecules in which the *probable* base sequence could be calculated knowing the relative abundance of the bases in the mixture and assuming (perhaps wrongly!) that they combine together at random. From such work it has been possible to deduce the base triplets associated with the coding of all the naturally occurring amino acids, but the *sequence* of bases in each triplet was not known. Using transfer RNA's, Nirenberg and Lender were able to determine the actual sequences and some of their results are given below

Alanine	GCU, GCC, GCA, GCG
Arginine	CGU, CGC, CGA, CGG, AGA, AGG
Aspartic acid	GAU, GAC
Glutamic acid	GAA, GAG
Glycine	GGU, GGC, GGA, GGG
Tryptophan	UGG

FAT METABOLISM

The principal function of fat is to provide a store of energy-rich materials which can be used to drive endergonic reactions during germination. The suitability of fat as an energy-producing food is shown by the following figures:

It will be seen that fat provides a superior means of storing energy than either carbohydrate or protein.

Apart from their importance as a storage material, they also, in

Complete oxidation of 1 gm of	Energy production	Water produced
Carbohydrate	4·2 kcal	0·55 gm
Fat	9·3 kcal	1·07 gm
Protein	5·6 kcal	0·41 gm

the form of phospholipids, are intimately involved in cytoplasmic structure.

In animal physiology reference is often made to fat metabolism and metabolic water as an adaptation to life in dry conditions. Very few similar references can be found in plant physiology, but one rather intriguing set of data is that of Haas and Hill (1935) who measured the fat content of some littoral Fucoids. They found that the fat content of *Pelvetia canaliculata* was almost twice that of *Ascophyllum nodosum* or *Fucus vesiculosus* and over four times that of *Himanthalia lorea*—in fact there was a close correlation between the period of exposure between tides and the fat content. Relatively little algal physiology has been done and it is interesting to speculate to what extent the exposed algae might use metabolic water, especially during neap tides when *Pelvetia* may be exposed for several days without any immersion.

The first stage in the utilization of fats is their catalytic hydrolysis by lipases. This probably takes place in three steps.

$$\text{Triglyceride} + H_2O \longrightarrow \text{Diglyceride} + \text{Fatty acid}$$
$$\text{Diglyceride} + H_2O \longrightarrow \text{Monoglyceride} + \text{Fatty acid}$$
$$\text{Monoglyceride} + H_2O \longrightarrow \text{Glycerol} + \text{Fatty acid}$$

The optimum pH varies according to the species from which the enzyme is extracted and the stage of development of the tissue. Thus Bamann showed that the optimum pH of unripe castor oil seeds is 8·5–10·5 and of ripe ungerminated seeds, 4·7.

The fates of the products of fat hydrolysis are shown in simple form in fig. 51.

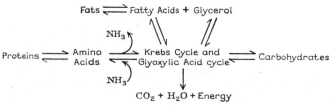

FIG. 51. Fates of products of fat hydrolysis

The main points to note are

(a) Glycerol may be used either in the formation of carbohydrate or in the production of energy via the Krebs' cycle.

(b) Fatty acids may also be used for the production of energy via the Krebs' cycle.

(c) Providing energy is supplied, fats may be resynthesized by a condensation of fatty acids and glycerol.

(d) The Krebs' cycle and its modification, the glyoxylic cycle, provide a means for the interconversion of carbohydrates, fats and protein.

The Fate of Glycerol

Glycerol enters the normal metabolic pathway as a result of its conversion to phosphoglyceraldehyde. This takes place in three main stages:

(i) A priming reaction with ATP to form α glycerol phosphate.

$$
\begin{array}{ccc}
CH_2OH & & CH_2OH \\
| & & | \\
CHOH + ATP & \rightarrow & CHOH \\
| & & | \\
CH_2OH & & CH_2O\,\boxed{P}
\end{array} \quad + ADP
$$

(ii) An oxidation (by dehydrogenation) coupled with cytochrome to form dihydroxyacetone phosphate.

$$
\begin{array}{ccc}
CH_2OH & & CH_2OH \\
| & & | \\
CHOH + Cyt. & \rightarrow & CO \\
| & & | \\
CH_2O\,\boxed{P} & & CH_2O\,\boxed{P}
\end{array} \quad + Cyt.\ H_2
$$

(iii) The formation of an equilibrium mixture, under the influence of triosephosphate isomerase, with phosphoglyceraldehyde.

$$
\begin{array}{ccc}
CH_2OH & & CHO \\
| & & | \\
CO & \rightleftharpoons & CHOH \\
| & & | \\
CH_2O\,\boxed{P} & & CH_2O\,\boxed{P}
\end{array}
$$

The phosphoglyceraldehyde can then enter into the reaction schemes outlined in chapter 1 (pages 28–30).

The Oxidation of Fatty Acids

It is usual to apply Greek letters to the C atoms of fatty acids, starting at the atom *next* to the COOH group, e.g.

$$
CH_3 \cdot \overset{\delta}{CH_2} \cdot \overset{\gamma}{CH_2} \cdot \overset{\beta}{CH_2} \cdot \overset{\alpha}{CH_2} \cdot COOH
$$

Two forms of fatty acid oxidation are possible. Oxidation of the α carbon atom was first discovered in 1952. It involves a relatively complex series of reactions in which glycollic acid ($CH_2OH \cdot COOH$) is oxidized to form H_2O_2. Details are not given but the overall reaction is

$$\overset{\beta}{R \cdot CH_2} \cdot \overset{\alpha}{CH_2} \cdot COOH + O_2 + NAD \rightarrow R \cdot CH_2 \cdot COOH + NAD\,H_2 + CO_2$$

The relative importance of this system and the β oxidation system, described below, is not known. Evidence suggests that the α oxidation is restricted to acids with 14–18 C atoms, so that possibly it serves to reinforce the β method in the early stages. It is possible that it serves also as a means of converting fatty acids with odd numbers of C atoms into those with an even number which could then undergo complete β oxidation.

The β oxidation is extremely complex and involves the participation of coenzyme A and ATP. A full account is not given here but the essentials of the system can be gained by studying Knoop's original suggestions, of alternate oxidation and hydrolysis of the β C atom, e.g.

$$R \cdot CH_2 \cdot CH_2 \cdot CH_2 \cdot CH_2 \cdot CH_2 \cdot CH_2 \cdot \overset{\beta}{CH_2} \cdot \overset{\alpha}{CH_2} \cdot COOH$$

$$\downarrow +[O]$$

$$R \cdot CH_2 \cdot CH_2 \cdot CH_2 \cdot CH_2 \cdot CH_2 \cdot CH_2 \cdot \overset{\beta}{CO} \cdot \overset{\alpha}{CH_2} \cdot COOH + H_2O$$

$$\downarrow +H \cdot OH$$

$$R \cdot CH_2 \cdot CH_2 \cdot CH_2 \cdot CH_2 \cdot \overset{\beta}{CH_2} \cdot \overset{\alpha}{CH_2} \cdot COOH + CH_3 \cdot COOH$$

$$\downarrow +[O]$$

$$R \cdot CH_2 \cdot CH_2 \cdot CH_2 \cdot CH_2 \cdot \overset{\beta}{CO} \cdot \overset{\alpha}{CH_2} \cdot COOH + H_2O$$

$$\downarrow +HOH$$

$$R \cdot CH_2 \cdot CH_2 \cdot \overset{\beta}{CH_2} \cdot \overset{\alpha}{CH_2} \cdot COOH + CH_3 \cdot COOH$$

The most important modification of this simple scheme is that instead of acetic acid, acetyl coenzyme A is formed

$$(CH_3 \cdot CO \sim S \cdot Co.A)$$

and this can enter the Krebs' cycle.

The Synthesis of Carbohydrates from Fatty Acids

Both carbohydrates and fatty acids enter the Krebs' cycle as acetyl coenzyme A, i.e.

Carbohydrates $\rightleftharpoons$ Pyruvic acid

Acetyl Co.A $\rightarrow$ Krebs' cycle

Fats $\rightleftharpoons$ Fatty acids

Whereas the reaction Fatty acid ⇌ Acetyl Co.A is reversible, the reaction Pyruvic acid → Acetyl Co.A is not. Therefore, although it is quite possible to synthesize fatty acids from carbohydrates by the route

Carbohydrate → Pyruvic acid → Acetyl Co.A → Fatty acids

the reverse route, as a means of synthesizing carbohydrates from fatty acids, is not possible.

A solution to the problem was advanced by Krebs in 1957—the glyoxylic acid cycle, which is, in effect, a modification of the original Krebs' cycle (see fig. 52).

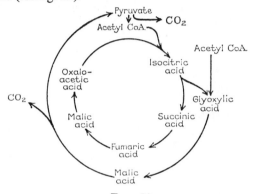

FIG. 52

The modification is that isocitric acid (formed in the normal way by the action of oxaloacetic acid and acetyl coenzyme A) is split into glyoxylic acid and succinic acid.

$$
\begin{array}{lll}
\text{CH}_2\cdot\text{COOH} & \text{CH}_2\cdot\text{COOH} \\
| & | \\
\text{HC}\cdot\text{COOH} & \rightarrow \quad \text{CH}_2\cdot\text{COOH} \quad + \quad \text{CHO} \\
| & | \\
\text{CHOH}\cdot\text{COOH} & \qquad\qquad\qquad\qquad\quad \text{COOH} \\
\text{Isocitric acid} & \quad\text{Succinic} \qquad \text{Glyoxylic} \\
& \quad\quad\text{acid} \qquad\quad \text{acid}
\end{array}
$$

The succinic acid can follow the normal Krebs' cycle but the glyoxylic acid combines with another molecule of acetyl Co.A to form malic acid

$$
\begin{array}{ll}
\text{COOH} \\
| \\
\text{CHO} & \qquad \text{COOH} \\
\quad + & \qquad | \\
& \qquad \text{CHOH} \\
\text{CH}_3 & \rightarrow \quad | \\
| & \qquad \text{CH}_2 \\
\text{CO} \quad + \text{HOH} & \qquad | \\
\wr & \qquad \text{COOH} + \text{Co.ASH} \\
\text{S}\cdot\text{Co}\cdot\text{A} & \qquad \text{Malic} \quad \text{Coenzyme} \\
& \qquad \text{acid} \qquad\quad \text{A}
\end{array}
$$

Malic acid can be converted into pyruvic acid by two possible routes:

(i) By oxidative decarboxylation (the Ochoa reaction).

$$\begin{array}{l} COOH \\ | \\ CH_2 \\ | \\ CHOH \\ | \\ COOH \end{array} + NADP \rightarrow \begin{array}{l} CH_3 \\ | \\ CO \\ | \\ COOH \end{array} + NADP\,H_2 + CO_2$$

(ii) By oxidation to oxaloacetic acid, followed by a reductive decarboxylation (Wood–Werkman reaction).

$$\begin{array}{l} COOH \\ | \\ CH_2 \\ | \\ CHOH \\ | \\ COOH \end{array} + NAD \rightarrow \begin{array}{l} COOH \\ | \\ CH_2 \\ | \\ CO \\ | \\ COOH \end{array} + NAD\,H_2$$

$$\begin{array}{l} COOH \\ | \\ CH_2 \\ | \\ CO \\ | \\ COOH \end{array} \rightarrow CO_2 + \begin{array}{l} CH_3 \\ | \\ CO \\ | \\ COOH \end{array}$$

The Synthesis of Fat from Carbohydrate

The reactions involved in the β oxidation of fatty acids to form acetyl Co.A are reversible. Therefore fatty acids can be synthesized from acetyl Co.A by the reverse series of reactions, and, of course, fats can be produced by the condensation of fatty acids and glycerol.

7

Mineral Nutrition

If an analysis is made of the ash resulting from the combustion of plant material, a wide range of inorganic elements can be detected. Among those commonly found may be mentioned sodium, potassium, calcium, manganese, magnesium, iron, phosphorus, sulphur, silicon, chlorine and aluminium, while other constituents, lost in the volatile products of combusion, include carbon, hydrogen, oxygen and nitrogen.

With the exception of carbon, hydrogen and oxygen, the only source of these elements available to the plant is as ions dissolved in the soil solution and absorbed through the root hairs. The fact that so many elements occur in the plant does not necessarily mean that they are all *essential*—their presence could well be explained by an inability of the absorbing mechanism to differentiate between ions which, although chemically closely related, differ widely in their utility as metabolites.

A guide to the basic chemical requirements in plant nutrition can be found in the results of water culture experiments. In these experiments plants—e.g. barley seedlings—are grown in jars containing a sterile solution of known chemical composition. The roots of the plant are allowed to dip into the solution and provision is made for aerating the solution by periodically blowing in a current of air. The outside of the container is normally covered with black paper to exclude light and so discourage algal growth. The formulae of the solutions used by Sachs are given below. It will be seen that by substituting various salts present in his complete control solution, he was able to study the effects of deficiency of one particular element.

On the basis of such experiments it was concluded that, provided the plant was supplied with H_2O and CO_2, only *seven* other elements were essential for normal plant growth. These elements were Ca, K, Fe, Mg, N (as nitrate), P (as phosphate) and S (as sulphate).

Subsequently it was found that this view of only ten elements being essential for plant growth was too restricted. By using very carefully controlled culture techniques with chemicals of an extremely high standard of purity, it has now been demonstrated that traces of other elements must also be present in the culture solutions. As examples of these trace elements mention may be made of manganese, cobalt, chromium, zinc, aluminium, molybdenum, copper and boron. The

	Complete solution	Minus Ca	Minus Fe	Minus N	Minus P	Minus S	Minus Mg	Minus K
$CaSO_4 \cdot 2H_2O$	0·25 gm	—	0·25 gm	0·25 gm	0·25 gm	—	0·25 gm	0·25 gm
$Ca(H_2PO_4)_2 \cdot H_2O$	0·25 gm	—	0·25 gm	0·25 gm	—	0·25 gm	0·25 gm	0·25 gm
$MgSO_4 \cdot 7H_2O$	0·25 gm	0·25 gm	0·25 gm	0·25 gm	0·25 gm	—	—	0·25 gm
$NaCl$	0·08 gm	0·08 gm	0·08 gm	0·08 gm	0·08 gm	0·08 gm	0·08 gm	0·08 gm
KNO_3	0·70 gm	0·70 gm	0·70 gm	—	0·70 gm	0·70 gm	0·70 gm	—
$FeCl_3 \cdot 6H_2O$	0·005 gm	0·005 gm	—	—	0·005 gm	0·005 gm	0·005 gm	0·005 gm
K_2SO_4	—	0·20 gm	—	0·52 gm	—	—	0·17 gm	—
$Na_2HPO_4 \cdot 12H_2O$	—	0·71 gm	—	—	—	—	—	—
KCl	—	—	—	—	—	—	—	—
$Ca(NO_3)_2$	—	—	—	—	0·16 gm	—	—	—
$CaCl_2$	—	—	—	—	—	0·16 gm	—	—
$MgCl_2$	—	—	—	—	—	0·21 gm	—	—
$NaNO_3$	—	—	—	—	—	—	—	0·59 gm
Distilled H_2O	1000 c.c.	1000 c.c.	1000 c.c.	1000 c.c.	1000 c.c.	1000 c.c.	1000 c.c.	1000 c.c.

TABLE 1. Details of Sach's formulae for water culture solutions

apparently successful results obtained by Sachs and his contemporaries can be attributed to the lower standard of purity of the chemicals available at that time.

It is customary to refer to the ten essential elements as the macronutrients and to those which need only be present in traces as the micronutrients or trace elements.

Mode of Action of the Mineral Nutrients

In general there are four main ways in which the inorganic nutrients can be utilized by the plant:

(a) By forming an integral part of complex organic molecules; this is especially true in the case of many macronutrients.

(b) By assisting in the functioning of enzyme systems, very probably by providing a means of 'binding' the enzyme and substrate to form an enzyme–substrate complex: most of the micronutrients appear to function in this way.

(c) By participating in an ionic balance, e.g. between mono- and bivalent cations, which is important for normal protoplasmic functioning, including the maintenance of differential membrane permeabilities.

(d) By providing an oxidation–reduction system in the case of variable valency elements, e.g. iron and copper.

Like the vitamins in animal nutrition, the mineral elements are often considered mainly in relation to the pathological symptoms produced by their deficiencies. This aspect is obviously of great importance to the plant pathologist, but from the physiological point of view it is much more important to consider the underlying mechanisms, especially when it is remembered that different plant species do not always show identical symptoms when faced with a deficiency of the same element. Conversely similar symptoms may result from deficiencies of several different elements. As an example of this the symptom of chlorosis may be considered.

Chlorosis implies a lack or absence of chlorophyll from the normally green parts of a plant and it can result from a deficiency of *any one* of the following elements: K, Fe, Ca, P, Mg, N, CO_2, H_2O, or a lack of light. Thus a shortage of any one of nine of the ten macronutrients can result in chlorosis although the chlorophyll molecule only consists of five different elements

$$\text{(e.g. chlorophyll a, } C_{55}H_{72}O_5N_4Mg)$$

A simplified scheme to show the various stages involved in chlorophyll synthesis is shown in fig. 53.

The necessity for iron in the synthesis of chlorophyll depends on the formation of iron–porphyrin complexes which are able to react

with magnesium to produce chlorophyll. Hence a deficiency of iron results in an inability to synthesize the necessary precursor. Since this complex is in its turn dependent on a supply of C, N, O and H elaborated by photosynthesis, then the necessity for an adequate supply of CO_2, H_2O, NO_3' and light is also explicable. Given all the above constituents in appropriate amounts, chlorosis may still occur. The key to this apparent paradox lies in the problem of the iron being

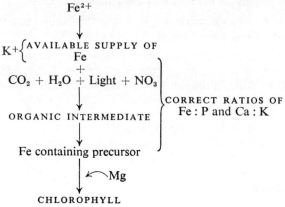

FIG. 53. Mineral elements involved in chlorophyll synthesis

present in a form suitable for incorporation into the organic inter-mediate. In order for this to take place it is necessary that potassium ions should be present and in a correct balance with the bivalent calcium ions, and that furthermore there should be a correct balance between the ions of iron and phosphorus (as phosphate).

Apart from this complex inter-relationship of many of the macro-nutrients in chlorophyll formation, there is a considerable body of information about the effect of individual elements.

Potassium. Possibly the most difficult element to understand is potassium. As far as is known it does not occur in combination in any organic compounds and its most characteristic distribution is in meristematic tissues, although it is rarely found to any significant extent in either the nucleus or the plastids.

Symptoms of potassium deficiency, apart from a tendency towards chlorosis, may include leaf 'scorch', due to a low water content in the leaves, reduced leaf production, abnormally shaped leaves (e.g. twisted) and also an increased amount of reducing sugars and other soluble carbohydrates present. There is also a marked increase in the concentration of amino acids and amides.

If the results of experiments carried out on plant and animal enzymes are considered together it will be found that there is an impressive array of enzymes which require the presence of K^+ ions for their activity—e.g.

(i) In the Krebs' cycle

$$\text{Succinate} \xrightarrow{\text{Succinic dehydrogenase}} \text{Fumarate}$$

$$\text{Malate} \xrightarrow{\text{Malic dehydrogenase}} \text{Oxaloacetate}$$

(ii) In Glycolysis

$$\text{Fructose 6 phosphate} + \text{ATP} \xrightarrow{\text{Phosphohexokinase}} \text{Fructose 1 : 6 diphosphate} + \text{ADP}$$

$$\text{Fructose 1 : 6 diphosphate} \xrightarrow{\text{Aldolase}} \text{Phosphoglyceraldehyde} + \text{dihydroxyacetone phosphate}$$

(iii) In more general terms it has also been found necessary for photophosphorylation, the incorporation of amino acids into ribosomes, the synthesis of starch from glucose and the synthesis of NAD.

If other cations—rubidium, ammonium, lithium or sodium are substituted for potassium in these enzyme systems, it will be found that in the case of Rb^+ and NH_4^+ the systems often show continued, but decreased, activity whereas it is rare to find any activity when Li^+ or Na^+ are substituted. A consideration of the radii of these ions in their hydrated state ($K^+ = 5 \cdot 3$ Å, $NH_4^+ = 5 \cdot 3$ Å, $Rb^+ = 5 \cdot 1$ Å, $Na^+ = 7 \cdot 9$ Å and $Li^+ = 10 \cdot 0$ Å) has led Evans and Sanger (1966) to point out the similarity in the dimensions of K^+, NH_4^+ and Rb^+ and to suggest that the mode of operation of K^+ is in some way to enter into the three dimensional structure of the protein-enzyme and so affect the configuration of the enzymes' reactive surfaces.

Calcium. The general effect of calcium deficiency is associated with the strength of the plant. If calcium is lacking, then the stems, particularly of herbaceous plants, tend to be weak and there is a poor degree of root development.

There are two suggestions as to the role played by calcium in the strength of the plant. One postulates that Ca^{2+} ions, combined with pectic acid as calcium pectate, form an essential part of the middle lamella of the cell wall, so that a shortage of calcium produces its effect by weakening the walls. Many investigators have analysed the components of the middle lamella and it would appear that in some cases pectic acid is absent. For example Conrad analysed fourteen different species of plants and found pectic acid to be present in only one.

Another explanation is concerned with the balance of Na^+, K^+ and Ca^{2+} ions. Maintenance of this balance within fairly narrow limits would appear to be essential for maintaining the differential permeability of the cell membranes. If this is the case then it is possible to trace a correlation between Ca^{2+} level, turgor and consequently rigidity of the stem.

A third suggested function of calcium is in the neutralization of organic acids. Studies on citrus fruits indicate that a calcium deficiency is associated with a high acid content, and attendant mottling, of the leaves. Whereas this may well be an important function of calcium in plants with a pronounced acid metabolism, it does not explain the role of calcium in all plants.

There is a considerable body of evidence to show that in calcium deficient plants there are extensive changes in cell division. In particular the number of cells undergoing normal mitosis in root tips is very much reduced and cells can be seen undergoing amitosis. A closer examination shows an abnormally high number of chromosome abnormalities present and it would appear that calcium ions play an important part in chromosome structure, perhaps in connecting the protein and DNA components.

Magnesium. The most obvious use of magnesium is in the formation of chlorophyll where, as mentioned earlier, it combines with an iron-containing precursor to form the actual chlorophyll molecule.

In addition, magnesium is intimately associated with phosphorus metabolism, particularly in the synthesis of ATP from ADP and inorganic phosphate. The mode of action of the magnesium ions is probably in serving to 'bind' the enzyme and substrate together and further details will be found in the section on micronutrients.

Iron. The function of iron in forming an iron–porphyrin chlorophyll precursor has already been discussed. Another group of iron-containing compounds are the cytochromes which play an essential role in the transfer of hydrogen from hydrogen acceptors (e.g. NAD and NADP) to molecular oxygen.

A deficiency of iron therefore, in addition to causing chlorosis, can seriously impair aerobic respiration and linked processes associated with it such as salt accumulation, details of which will be found in chapter 8.

Phosphorus. Apart from the necessity of a suitable Fe : P ratio for the synthesis of the chlorophyll precursor there are two other important ways in which a deficiency of phosphorus can affect a plant's metabolism.

The wide occurrence of phosphorus-containing lipoids and proteins results in a general feeble growth of the plant under conditions of phosphorus shortage, while the use of phosphorus-containing compounds as a means of transferring energy within living organisms— i.e. by the synthesis of ATP from ADP and inorganic phosphate, results in a serious impairment of most of the synthetic reactions

| Element | Chlorosis | Other leaf effects | Root growth | Stem growth | Rates of | | |
					Photo-synthesis	Protein synthesis	Respiration
K	May occur	Scorching, bronzing, twisting, reduced numbers			Inhibition	Retardation of early stages. ∴ Amide accumulation	Acceleration
Ca	May occur in conjunction with K	Mottling (in citrus)	Definite retardation	Reduced vigour. Sometimes feeble stature		Inability to utilize NO_3^-	
Fe	Very marked		Reduced	Reduced	Reduced	Reduced	Reduced
S	Some		Extensive	Hard and woody		Reduced ∴ Accumulation of nitrate	
Mg	Very marked	Reduced growth	Reduced	Reduced	Reduced		
N	Very marked	Reduced growth	Reduced	Reduced	Reduced	Reduced	Reduced
P	May occur in conjuction with Fe: P		Feeble	Feeble	Reduced	Reduced	Reduced

TABLE 2. Some effects of deficiencies of the seven macronutrient elements

occurring in the plant unless the supply of phosphate is adequate. Associated with this is the tendency of phosphorus-deficient plants to accumulate carbohydrates and soluble nitrogenous compounds.

Sulphur and Nitrogen. The effects of these two elements can be considered together since they both exert their effects by entering into the formation of amino acids and proteins. In the case of nitrogen, its occurrence in amino acids is of course ubiquitous, and, in the absence of an external supply of nitrogen, growth soon ceases.

The occurrence of sulphur as a constituent of amino acids is not so general, but its occurrence in the amino acid cystine is very important. Symptoms of sulphur deficiency in tomato plants include yellow-green leaves, an extensive root system, a hard woody stem and some accumulation of nitrates and carbohydrates. In the case of legumes there may be an increase in the number of root nodule bacteria if sulphur is lacking.

A summary of the major effects of macronutrient deficiencies is given in table 2.

The Micronutrient (trace) Elements

The individual symptoms arising from deficiencies of particular trace elements are too numerous to discuss in a book of this type.

Rather, an attempt will be made to consider the mode of action of these micronutrients.

Unlike the macronutrients, it is not possible to give a full total of micronutrients although the work of Arnon yields some interesting results. He grew several different species of plants—barley, asparagus and lettuce—in culture solutions containing increasing numbers of elements, e.g.

1st solution contained K Ca PO_4''' NO_3' Fe Mg SO_4''

2nd   ,,   ,,   elements of 1st solution
                           + B Zn Mn Cu Cl'

3rd   ,,   ,,   elements of 2nd solution
                           + Mo Ti V Cr W Co Ni

4th   ,,   ,,   elements of 3rd solution
                           + Al As Cd Sr Hg Pb Li Rb Br I F Se Be Na

His results for lettuce plants were

Solution	Average fresh wt in gm	
	Shoots	Roots
1	71·4	14·5
2	105·7	22·0
3	1068·3	188·6
4	984·4	196·2

If these results are considered it will be seen that

(a) The addition of the elements B Zn Mn Cu and Cl to the solution of the seven macronutrients results in approximately a 50% increase in the fresh weight of both shoot and roots. Of these five elements, two have been mentioned in connexion with enzyme reactions.

(b) The addition of the further elements of solution 3 results in an even more spectacular increase in fresh weight, of the order of nine- to tenfold. In view of the importance of Mo for nitrate reduction and Cr, Co (and possibly Ni) as participants in other enzyme reactions, this can therefore be explained but it still leaves the possibility that titanium, vanadium and tungsten may play a part, although there is little evidence that this is the case.

(c) The addition of the further elements of the fourth solution makes little difference to the growth of the plant, suggesting that few if any of these elements are essential. Aluminium may participate in some enzyme systems but in general its effects are often toxic.

In terms of the number of elements involved in plant growth, apart from C, O and H, there are seven macronutrients and about ten micronutrients (B Zn Mn Cu Mo Cr Co Ni Al Na). In general it may be stated that the micronutrients (and a few of the macronutrients which will further be considered in this section) play their part by participation in enzyme systems.

McElroy and Nason (1954) in their comprehensive review of the subject suggest that there are two basic ways in which this may be brought about.

(i) Particularly in the case of iron (a macronutrient), copper and molybdenum, the essential feature is an ability to act as electron carriers by virtue of valency changes. Thus in the case of iron acting in the cytochrome/cytochrome oxidase system, the oxidation of a reduced hydrogen acceptor involves the production of a hydrogen ion and an electron, the latter being used in the conversion of ferric iron to ferrous.

$$AH_2 \rightarrow A + 2H^+ + 2e$$
$$2e + 2Fe^{3+} \rightarrow 2Fe^{2+}$$

The ferrous ions are then reoxidized at the expense of atmospheric oxygen:

$$2Fe^{2+} + \tfrac{1}{2}O_2 + 2H^+ \rightarrow 2Fe^{3+} + H \cdot OH$$

Copper reacts in a similar way in the ascorbic acid/ascorbic oxidase system:

$$AH_2 \rightarrow A + 2H^+ + 2e$$
$$2Cu^{2+} + 2e \rightarrow 2Cu^+$$
$$2Cu^+ + \tfrac{1}{2}O_2 + 2H^+ \rightarrow 2Cu^{2+} + H \cdot OH$$

The cyclical nature of these reactions is shown in the case of iron in fig. 54.

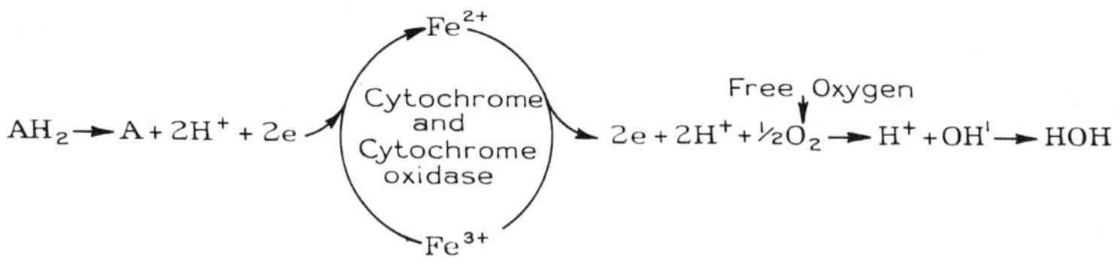

FIG. 54. To show how iron valency changes are involved in the cytochrome mediated reaction $AH_2 + \tfrac{1}{2}O_2 \rightarrow A + H_2O$

In the case of molybdenum, the metal acts as an electron transfer system for the reduction of nitrate to nitrite. It is probable that it also acts by valency changes (e.g. when involved in nitrate reduction of micro-organisms):

$$2H + 2Mo^{6+} \rightarrow 2Mo^{5+} + 2H^+$$
$$2Mo^{5+} + 2H^+ + NO_3' \rightarrow 2Mo^{6+} + NO_2' + H_2O$$

(ii) The other important method by which micronutrients (and also the macronutrient, magnesium) act is by combining with the enzyme substrate complex. The actual details of the chemical reactions will not be considered here. Suffice it to say that under this heading there are several ways in which the inorganic ions may act, viz.

(a) By forming a compound with the substrate, which enables the enzyme to act more easily.

(b) By forming a compound with the enzyme and accelerating the formation of intermediate compounds.

(c) By taking part in the formation of a carrier when group transfer, e.g. phosphate, is being catalysed.

No mention has so far been made of boron. It differs from the other micronutrients in that there is no evidence to suggest a connexion with enzyme systems and it also differs in that it is absorbed as an anion, i.e. borate or tetraborate, rather than as a cation like all the other metallic nutrients. Its mode of action is obscure and incompletely known and at the present stage there is little point in more than summarizing some of the results of deficiencies. These include

(a) Accelerated mitoses at the growing points, but with retarded cell differentiation.

(b) High amino acid content, with an associated lack of protein.

(c) Carbohydrate accumulation.

(d) Inhibited pollen tube development.

(e) Decreased phloem transport (e.g. of sugar).

Sodium has been reported as necessary for the growth of halophytes and Brownell (1965) has described the effects of sodium deficiency on the halophyte *Atriplex vesicaria*—in the absence of sodium there was a marked retardation of growth, extensive chlorosis and finally death. A particularly interesting feature, in view of Evans and Sorgers' hypothesis for the mode of action of potassium, is that in the case of *Atriplex*, sodium cannot be replaced by potassium, implying that the halophyte's enzymes react with ions of a larger diameter.

8

Respiration

Introduction

The term respiration was first used by animal physiologists to describe the breathing movements of animals, but subsequently was extended to include the transport of oxygen to the cells and also the chemical reactions by which organic compounds (usually hexose sugars) were combined with oxygen to produce carbon dioxide, water and energy, the latter being utilized whenever the organism carried out work of any kind.

In the case of plants the problem is complicated because

(*a*) Breathing movements are not performed.
(*b*) The gaseous exchange typical of animals is often masked by photosynthesis.
(*c*) Oxygen need not be present.

Although there is considerable diversity in the definitions employed by physiologists, the following will be found to agree with all but the most extreme views.

External Respiration. In the green plant the process by which oxygen is brought to the respiratory centres (usually the mito-chondria) of the cell may vary in extent from mere diffusion of oxygen from the chloroplasts to the mitochondria if the plant is illuminated, to diffusion from the outside air, through the stomata, across the intercellular spaces and through the cell wall if the plant is in darkness. (In mammals, on the other hand, it would include the passage of oxygen down the trachea, bronchi and bronchioles, diffusion across the alveolar epithelium, transport as oxyhaemoglobin and subsequent diffusion to the cells.)

Internal Aerobic Respiration. The oxidation of cellular food material, utilizing molecular oxygen, with the consequent production of energy, is known as internal aerobic respiration. The substrate is usually a hexose sugar and the reaction may be expressed simply as

$$C_6H_{12}O_6 + 6O_2 \longrightarrow 6CO_2 + 6H_2O + Energy$$

This process is sometimes referred to as tissue respiration.

The site of internal respiration is the mitochondria. These are of

133

varying shape, varying from ovoid-spheroid to rod-like. Electron microscope examination shows that they are made up of a pair of external membranes. The innermost of these has branches at right angles which pass towards the interior. Dehydrogenases, oxidases and ATP are situated between the two external membranes.

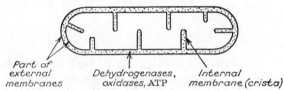

Part of external membranes   Dehydrogenases, oxidases, ATP   Internal membrane (crista)

FIG. 55. Diagrammatic representation of mitochondrion in longitudinal section

Internal Anaerobic Respiration. The production of energy from cellular food material *in the absence of molecular oxygen* is known as internal anaerobic respiration. Many higher plants are able to carry it out for considerable periods of time if the oxygen concentration of their environment falls to a low level, but it is doubtful if many animals are capable of this, although parts of them may carry out forms of anaerobic respiration over short periods (e.g. the production of lactic acid by muscle).

Fermentation is the form of anaerobic respiration carried out by some fungi and bacteria. The modern tendency is to use the term fermentation also for the anaerobic respiration of higher plants.

It will be seen that, apart from external respiration, which is really a transport mechanism, respiration can be considered as a process by which cells obtain their supplies of energy by the degradation of complex compounds.

Fermentation

The reactions involved have already been outlined in chapter 1 (pages 28–30) and consist of the following stages:

(i) The raising of the free energy level of the substrate by two transphosphorylations from ATP.

Glucose + ATP $\longrightarrow$ Glucose 6 phosphate + ADP
Glucose 6 phosphate $\longrightarrow$ Fructose 6 phosphate
(an isomerization)
Fructose 6 phosphate + ATP $\longrightarrow$ Fructose 1:6 diphosphate + ADP

viz. $C_6H_{12}O_6 + 2ATP \longrightarrow C_6H_{10}O_6-2\!\left(P\right) + 2ADP$

(ii) A splitting of the fructose 1: 6 diphosphate into two molecules of triose phosphate under the influence of the adding enzyme aldolase

Fructose 1:6 diphosphate   →   Phosphoglyceraldehyde +
                                                            Dihydroxyacetone phosphate

Only the phosphoglyceraldehyde is utilized and the conversion of dihydroxyacetone phosphate into phosphoglyceraldehyde is catalysed by the enzyme triose phosphate isomerase.

$$
\begin{array}{ccc}
CH_2OH & & CHO \\
| & & | \\
CO & \rightarrow & CHOH \\
| & & | \\
CH_2O\,\boxed{P} & & CH_2O\,\boxed{P}
\end{array}
$$

(iii) The oxidation of the phosphoglyceraldehyde by the removal of hydrogen to coenzyme I (NAD). This reaction involves a preliminary hydrolysis

$$
\begin{array}{ccc}
CHO & & \overset{\displaystyle H}{\underset{\displaystyle \diagdown OH}{C\diagup{-}OH}} \\
| & & | \\
CHOH & +\ H\ OH\ \rightarrow & CHOH \\
| & & | \\
CH_2O\,\boxed{P} & & CH_2O\,\boxed{P}
\end{array}
$$

The unstable hydrate, bound to the dehydrogenase enzyme, is then dehydrogenated. Phosphoglyceric acid is *not* formed in the free state at this stage but is in the form of a phosphoglycerate–enzyme complex. Separation of enzyme from this complex is achieved by a second phosphorylation (utilizing inorganic phosphate) but the energy changes are such as to result in the terminal phosphate radical in the diphosphoglyceric acid being produced at a high energy level.

The overall reaction is

$$
\begin{array}{ccc}
\overset{\displaystyle H}{\underset{\displaystyle \diagdown OH}{C\diagup{-}OH}} & & \\
| & & COO\,\boxed{P} \\
CHOH & +\ NAD + HO\,\boxed{P}\ \rightarrow & CHOH \quad + NAD\ H_2 + H_2O \\
| & & | \\
CH_2O\,\boxed{P} & & CH_2O\,\boxed{P}
\end{array}
$$

                                1:3 Diphospho-
                                glyceric acid

(iv) The diphosphoglyceric acid undergoes a series of reactions by which the high energy phosphate group is transferred to ADP. This is followed by a substrate phosphorylation and the formation of pyruvic acid. The overall reaction is:

$$\begin{array}{cccc}
COO\,\text{\textcircled{P}} & & COOH & \\
| & & | & \\
CHOH & + 2ADP \rightarrow & CO & + H_2O + 2ATP \\
| & & | & \\
CH_2O\,\text{\textcircled{P}} & & CH_3 & \\
& & \text{Pyruvic acid} &
\end{array}$$

Since *two* molecules of phosphoglyceraldehyde are produced for each molecule of hexose used, there will be a production of *four* molecules of ATP. Since two were used in the initial priming of hexose, there is a nett production of 2 molecules of ATP per molecule of glucose used.

(v) The reoxidation of the NAD H_2 is brought about by the coupled reduction of acetaldehyde. Pyruvic acid is first decarboxylated

$$CH_3 \cdot CO \cdot COOH \rightarrow CH_3 \cdot CHO + CO_2$$
$$NAD\ H_2 + CH_3CHO \rightarrow CH_3 \cdot CH_2OH + NAD$$

The essential stages are summarized in fig. 56.

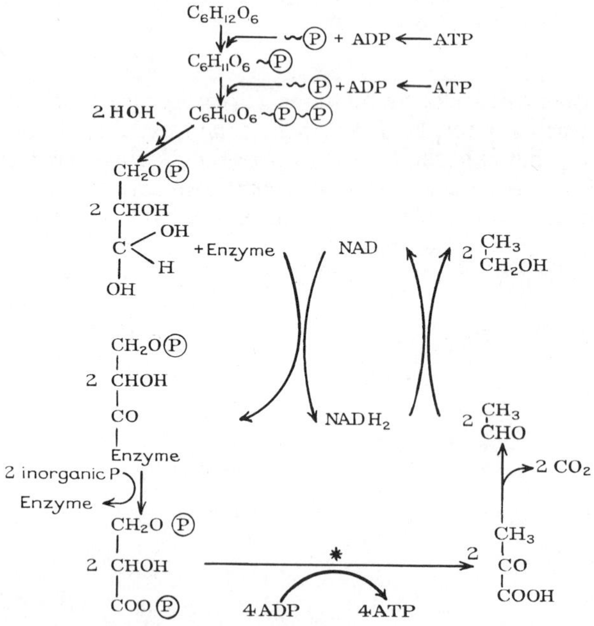

FIG. 56

* The intermediate compounds formed are phosphoglyceric acid, enol phosphopyruvic acid and phosphopyruvic acid.

Little mention has been made of the enzymes involved in fermentation. In 1897 H. and E. Buchner accidentally discovered that it was possible to extract a juice from yeast cells which was able, in the absence of living cells, to ferment sugar. This extract showed the properties of an enzyme and was called 'zymase'. Subsequent work has demonstrated that zymase consists of a number of separate enzymes, some of which are listed in table 3, together with the reactions which they catalyse.

Enzyme	Reaction
Hexokinase	Glucose + ATP → Glucopyranose phosphate + ADP
Phosphohexoisomerase	Glucopyranose phosphate → Fructofuranose 6 phosphate
Phosphohexokinase	Fructose 6 phosphate + ATP → Fructofuranose 1:6 diphosphate + ADP
Aldolase	Fructofuranose 1:6 diphosphate → Phosphoglyceraldehyde + Dihydroxyacetone phosphate
Phosphotriose isomerase	Dihydroxyacetone phosphate ⇌ Phosphoglyceraldehyde
Triose phosphate dehydrogenase	Phosphoglyceraldehyde + NAD + HO$\bigcirc\!\!\!P$ → Diphosphoglyceric acid + NAD H$_2$
Carboxylase	Pyruvic acid → Acetaldehyde + CO$_2$
Alcohol dehydrogenase	Acetaldehyde + NAD H$_2$ → Ethyl alcohol + NAD

TABLE 3.   The main constituents of zymase

Anaerobic Respiration

When a plant is placed in an atmosphere lacking oxygen it will, for a varying length of time, continue to produce CO_2 and also, very often, will form ethyl alcohol. This suggests an obvious connexion with fermentation and there are many facts which, when considered together, form a formidable body of evidence in support of the thesis that anaerobic respiration and fermentation are basically similar, if not identical, processes. The evidence is considered below under four headings.

The Ratios Hexose: (Alcohol + CO$_2$) and Alcohol: CO$_2$. If anaerobic respiration can be represented by the equation for fermentation

$$C_6H_{12}O_6 \longrightarrow 2C_2H_5OH + 2CO_2$$

it should be possible to construct a 'balance sheet' in terms of the loss of hexose substrate and the production of ethyl alcohol and CO_2.

E.g. if in a given period of time,

> a gm of hexose are consumed
> b gm of ethyl alcohol are produced

and c gm of carbon dioxide are produced

then, if anaerobic respiration proceeds by the same path as fermentation

$$a = b + c$$

Furthermore the ratio $\dfrac{\text{Wt of alcohol produced}}{\text{Wt of CO}_2 \text{ produced}} = \dfrac{b}{a}$ should be in

the same proportion as

$$\frac{\text{Molecular wt of alcohol}}{\text{Molecular wt of CO}_2} = \frac{46}{44} = 1.04$$

Unfortunately there are not many results available of experiments of this type. Nabokich in 1903 worked with pea seedlings and found a very close approximation between a and $b + c$. In general he found that $b + c$ was about 2% less than a, which is well within the

limits of experimental error. His values for the ratio $\dfrac{\text{Wt of C}_2\text{H}_5\text{OH}}{\text{Wt of CO}_2}$

ranged from 1.03 to 0.96 which again give further support for the identity of fermentation and anaerobic respiration.

More recent work by Fidler (1948–1951) with apples gives results for the Hexose : CO_2 + Ethyl alcohol ratio which again support the identity of the two processes.

Stiles and Leach summarize the results of many determinations for the ratio C_2H_5OH : CO_2. Agreement with the theoretical value of 1.04 is by no means general and gives a variation from 0.00 to 1.02 for different species under anaerobic conditions. The 0.00 values were obtained with potatoes and here lactic acid is produced instead of alcohol although the possibility cannot be ruled out that alcohol is first produced and then further metabolized. It is interesting to notice that if enzymes are extracted from pressed tissues they usually convert sugar to alcohol and CO_2 in such a way that the value for C_2H_5OH : CO_2 is approximately one even in the case of plants such as the potato which, when intact, show values widely different from the theoretical.

The Enzymes Involved. The extraction of zymase from crushed plant tissues was first claimed in 1903, when zymase was considered to be a single enzyme. The vast amount of work since that date has since shown that all the enzymes which make up the zymase complex in yeast (see page 137) are also present in higher plants, and the various reactions which they individually catalyse also take place.

The essential reactions of glycolysis in particular seem to be common to both anaerobic respiration and fermentation.

The Effect of Phosphate. If extracted yeast juice is allowed to ferment sugar, then the rate of fermentation can be increased by the addition of inorganic phosphate. This can readily be explained since the phosphate is presumably being utilized in the synthesis of ATP. If an extract of higher plants is placed with a sugar solution, then phosphate can again be shown to have an acceleratory effect on CO_2 production and, often, on alcohol production.

The Formation of Acetaldehyde. Acetaldehyde was shown to be an intermediate in fermentation by Neuberg. He found that if sulphite was added to a fermenting liquor then acetaldehyde sulphite was formed and tended to accumulate, thus blocking alcohol formation. Later Neuberg was able to produce a similar blockage with anaerobically respiring pea seedlings.

Although there is no single piece of evidence which conclusively shows that fermentation and anaerobic respiration are identical processes, the sum of all the evidence builds up a formidable case for considering that they are at least broadly similar, but that some of the terminal reactions of anaerobic respiration may differ slightly from those of fermentation.

Before considering aerobic respiration, it is well worth considering the ecological significance of anaerobic respiration.

In general anaerobic conditions are the exception rather than the rule. Perhaps the commonest anaerobic conditions are associated with flooding of the soil and the consequent removal of air from the soil spaces. Even under these conditions of external anaerobiosis many plants survive, not by virtue of their ability to carry out anaerobic respiration but by their ability to overcome an external oxygen shortage by internal ventilation through the intercellular spaces.

The other common condition where anaerobic respiration may occur is in dormant seeds, especially those with a hard and impervious testa, but here the rate of respiration is very low. During the early stages of germination there may also be a short period of anaerobiosis, especially if there is a higher water content in the surrounding soil.

A higher plant with very pronounced tendencies towards anaerobic respiration is rice, and in this case germination actually seems to be favoured by an absence of oxygen.

The energy produced in anaerobic respiration is more than adequate for the plant's needs, but vital processes such as salt absorption and translocation, which require energy, fail rapidly unless oxygen is present, suggesting that it is not possible to couple the anaerobic energy source to the dependent processes.

Aerobic Respiration

With the possible exception of *Elodea Canadensis*, all higher plants which have been investigated respire anaerobically if the external oxygen supply is removed. It is therefore reasonable to consider the possibility that aerobic and anaerobic respiration both share a common series of reactions but that the final products differ according to whether or not oxygen is available. Further, it would seem not unlikely that the difference between the two processes might well reside in the manner by which the reduced coenzymes are reoxidized. Before this latter speculation can be considered in detail, the evidence for a 'common path' must first be examined more closely.

As far as is known the enzyme complex zymase is present in the tissues of all higher plants and its action is unaffected by the presence of oxygen. Possibly then the first stage in aerobic respiration takes place under the influence of zymase so that the common path theory could be represented as

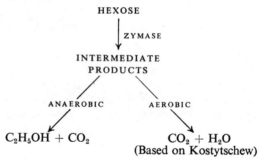

HEXOSE

ZYMASE

INTERMEDIATE
PRODUCTS

ANAEROBIC   AEROBIC

$C_2H_5OH + CO_2$ $CO_2 + H_2O$
(Based on Kostytschew)

Bearing this scheme in mind it is possible to consider the evidence for it under three main headings

The Effect of a Short Anaerobic Period on a Following Aerobic Period. When a plant is transferred from aerobic to anaerobic conditions for a few hours and then returned to aerobic conditions, it is found that the rate of respiration, as measured by the CO_2 output, is temporarily increased on its return to the aerobic environment, i.e.

If initial rate of aerobic respiration $= x$ mgm CO_2 per hour
and final rate of aerobic respiration $= y$ mgm CO_2 per hour
then temporarily $y > x$.

The most likely explanation of this is that under anaerobic conditions there is an accumulation of intermediate products which are easily oxidized on returning the plant to air.

The Effect of Temperature. Over the temperature range of 0°–35° C the Q_{10} for aerobic respiration is approximately two (see page 9). A similar value obtains for anaerobic respiration over the same range. Although this does not demonstrate the fact of a common pathway, it does suggest that a series of similar reactions are common to the two processes.

The Effect of Inhibition with HCN.* The mode of action of cyanide on tissues depends on its concentration.

At low concentrations, i.e. not greater than M/500, it reacts with metals—e.g. Fe and Cu—and so effectively blocks their participation in cytochrome/cytochrome oxidase and ascorbic acid/ascorbic oxidase systems respectively.

At higher concentrations, i.e. not less than M/100, it forms cyanhydrins with aldehydes and ketones.

The effect of high and low concentrations of cyanide on aerobic and anaerobic respiration is summarized below

Concentration of cyanide	Aerobic respiration	Anaerobic respiration
M/500	Inhibited	No effect
M/100	Inhibited	Inhibited

These results are consistent with the idea that there is a stage common to both aerobic and anaerobic respiration (viz. the formation of triose phosphate in glycolysis) which is sensitive to M/100 HCN but that there is a stage involving metallic oxidases, sensitive to M/500 HCN, which is peculiar to aerobic respiration.

The case for a stage common to both aerobic and anaerobic respiration therefore rests on

(a) The universal occurrence of aerobic and anaerobic respiration (with the possible exception of *Elodea*).
(b) The universal occurrence of enzymes making up the zymase complex.
(c) The accelerated rate of aerobic respiration following a short period of anaerobiosis.
(d) The similar effects of temperature.
(e) The effects of inhibition with HCN.

Taken together, this provides a convincing body of evidence.

* Not *all* tissues have their respiration easily inhibited by cyanide, but this may well be due to the presence of excess cytochrome oxidase. It does not invalidate the strength of the argument considered here.

Oxidation in Aerobic Respiration

The usual equation for aerobic respiration is

$$C_6H_{12}O_6 + 6O_2 \rightarrow 6CO_2 + 6H_2O + 674 \text{ cals}$$

This is the reverse of the simple equation given for photosynthesis on page 83. Like that equation it is useful in serving as a quantitative summary for the overall reaction but it fails to give a complete picture in that it implies that only one reaction is involved and that there is a direct oxidation of hexose by molecular oxygen.

It has been mentioned previously (page 20) that such direct oxidations are of rare occurrence in biological reactions and that removal of hydrogen is a much more usual step.

Palladin, in a series of papers from 1908 to 1912, suggests a modification of this equation to account for a dehydrogenation.

In his scheme he includes R to represent 'chromogens' or hydrogen acceptors which receive hydrogen partly from water and partly from the carbohydrate before being reoxidized by molecular oxygen.

$$C_6H_{12}O_6 + 6H_2O \rightarrow 6CO_2 + 12H_2$$
$$12H_2 + 12R \rightarrow 12RH_2$$
$$12RH_2 + 6O_2 \rightarrow 12H_2O$$

On addition $\quad C_6H_{12}O_6 + 6O_2 + 6H_2O \rightarrow 6CO_2 + 12H_2O$

It is interesting to notice that this is the reverse of the alternative equation for photosynthesis, described on page 86.

The mechanism of oxidation in aerobic respiration is principally concerned with utilizing molecular oxygen to restore the coenzyme reduced in the initial stages of glycolysis. There are several oxidases which are available, in particular cytochrome oxidase and ascorbic oxidase. Although it is not yet possible to generalize, it is true to say that modern work increasingly favours the cytochromes as the major oxidases involved although in a few cases the ascorbic system is utilized.

Details of the functioning of the cytochrome system have already been given (page 22). The important feature is that it provides a means of using molecular oxygen to reoxidize reduced coenzyme I according to the scheme:

$P + NAD + \text{Phosphoglyceraldehyde} \rightarrow \text{Diphosphoglyceric acid} + NADH_2$

$NAD\,H_2 + 2Fe^{3+} \text{ cytochrome} \rightarrow 2Fe^{2+} \text{ cytochrome} + 2H^+ + NAD$

$2Fe^{2+} \text{ cytochrome} + 2H^+ + \tfrac{1}{2}O_2 \rightarrow 2Fe^{3+} \text{ cytochrome} + H\cdot OH$

Molecular
oxygen

The main consequences are

(a) The ready oxidation of NAD H_2 without the formation of ethyl alcohol from pyruvic acid.

(b) The availability of the pyruvic acid to participate in the Krebs' cycle.

(c) The ability of the Krebs' cycle to operate (it is linked with the cytochrome system).

(d) The production of much more energy for each gram molecule of hexose utilized—each turn of the Krebs' cycle produces 15 ATP molecules, and there are two turns for each molecule of hexose consumed. The total nett gain is in the order of 38 ATP molecules, each representing 8,000–9,000 cal.

(e) The Krebs' cycle opens up the possibility of pyruvic acid being diverted into other metabolic pathways.

A summary of the inter-relationships of fermentation and aerobic respiration is given in fig. 57.

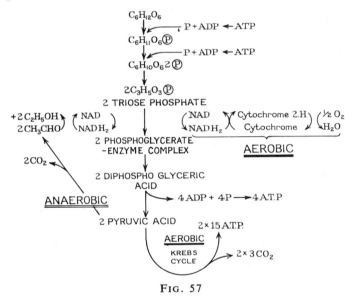

FIG. 57

Respiration and Photosynthesis

It is axiomatic that the energy released in respiration represents solar energy fixed by photosynthesis, but it is only recently that it has become apparent just how similar are the two processes from a chemical point of view (see fig. 58).

The essential light stage in photosynthesis, in which NADP is reduced by hydrogen from water and ATP is synthesized, is unique. If however the fate of the NADP H_2 is traced it will be found that the same metabolic pathways are traversed, but in reverse.

Thus the NADP H_2 in photosynthesis is reoxidized in the reaction phosphoglyceric acid $\rightarrow$ triose phosphates which condense to form

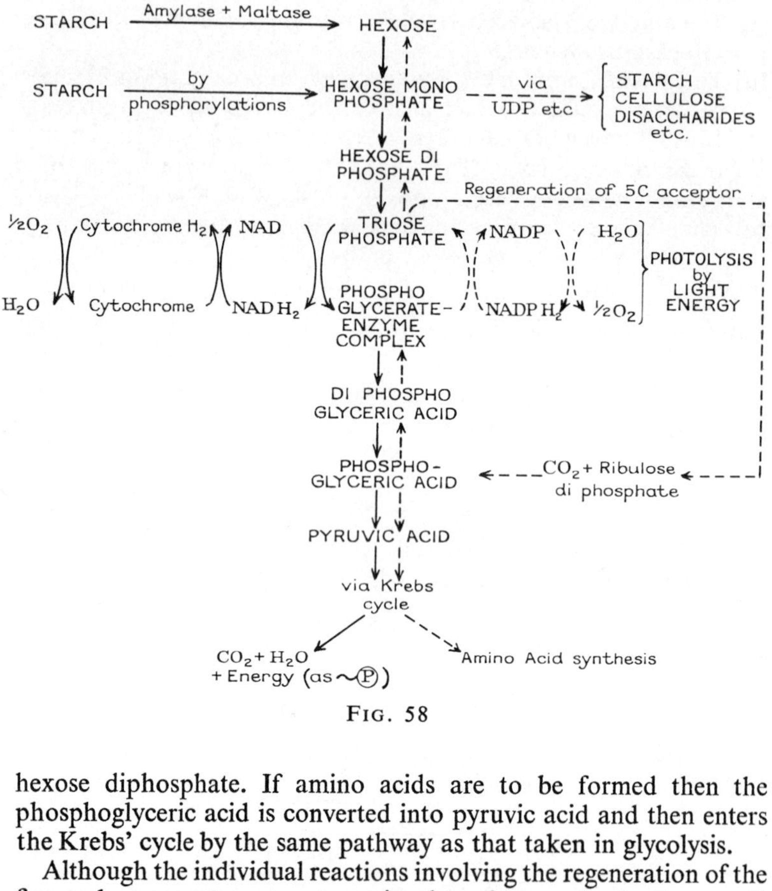

FIG. 58

hexose diphosphate. If amino acids are to be formed then the phosphoglyceric acid is converted into pyruvic acid and then enters the Krebs' cycle by the same pathway as that taken in glycolysis.

Although the individual reactions involving the regeneration of the five carbon acceptor are not restricted to photosynthesis, the overall balance which is achieved may perhaps be peculiar to that process.

The Rate of Respiration

Expression of Results: The Respiratory Quotient. From the simple equation $C_6H_{12}O_6 + 6O_2 \rightarrow 6CO_2 + 6H_2O$ it can be calculated that the volume of oxygen consumed is equal to the volume of CO_2 produced.

The ratio $\dfrac{\text{Volume of } CO_2 \text{ produced}}{\text{Volume of } O_2 \text{ consumed}}$ (in the same time) is known

as the respiratory quotient and is usually written as R.Q.

The simplest method of measuring the R.Q. is to use manometric methods, as described on page 37. Two sets of manometers are required: from those containing KOH it is possible to measure the oxygen consumption and from the other, without KOH, the *difference* between O_2 consumption and CO_2 consumption is obtained. Obviously if there are no volume changes recorded in the second series of manometers, the R.Q. = 1. If there is a volume change, then it is necessary to follow the method of calculation shown on page 37. This version of the apparatus is not very sensitive and the volume changes caused by the different solubilities of oxygen and carbon dioxide can be ignored.

The value of the R.Q. lies in its ability to give some idea as to the type of substrate being used and of the degree of aerobic respiration proceeding.

In the case of a substrate of hexoses, then R.Q. = 1.

If, instead of hexoses, fats are respired, then the value of the R.Q. falls.

E.g. in the case of the fat tripalmitin

$$CH_2OOC\cdot C_{15}H_{31}$$
$$|$$
$$CHOOC\cdot C_{15}H_{31}$$
$$|$$
$$CH_2OOC\cdot C_{15}H_{31}$$

the equation for its respiration can be written as

$$2C_{51}H_{98}O_6 + 145O_2 \longrightarrow 102CO_2 + 98H_2O$$

By applying Gay Lussac's law, the R.Q. $= \dfrac{102}{145} = 0\cdot70.$

In the case of protein oxidation either the R.Q. $\simeq 1\cdot0$ or $\simeq 0\cdot80$. In the former case ammonia is also produced, so that this value is unlikely to be confused with carbohydrate oxidation. In the lower value, amides are produced; when they are subsequently oxidized the R.Q. rises above one.

Other possible substrates are organic acids. The equation for the respiration of malic acid is

$$C_4H_6O_5 + 3O_2 \longrightarrow 4CO_2 + 3H_2O$$
$$\therefore R.Q. = \frac{4}{3} = 1\cdot33$$

Alternatively malic acid may be first converted into hexose

$$2C_4H_6O_5 \rightarrow C_6H_{12}O_6 + 2CO_2 *$$

If this hexose is then only partly respired it is possible that the nett R.Q. is considerably greater than 1·3.

Values of R.Q.'s in the region 0·2–0·3 are sometimes found and these are often associated with a combination of hexose respiration and organic acid synthesis at the expense of CO_2, i.e.

$$C_6H_{12}O_6 + 2CO_2 \rightarrow 2C_4H_6O_5$$

Similar low values are obtained in stages of the germination of seeds containing abundant fat reserves and this is attributed to the combined effects of the seed using a fat substrate for respiration and also synthesizing carbohydrate from fat.

At low oxygen concentrations, anaerobic respiration will occur in addition to some aerobic which increases the amount of CO_2 production and so raises the value of the R.Q.

A summary of approximate R.Q.'s and possible substrates is given in table 4.

R.Q.	Substrate
>1·0	Carbohydrate with some anaerobic respiration Carbohydrate synthesized from organic acids Organic acids
1·0	Carbohydrates
0·99	Proteins with NH_3 formation
0·8	Proteins with amide formation
0·7	Fats, e.g. tripalmitin
0·5	Fats with associated carbohydrate synthesis
0·3	Carbohydrates with associated organic acid synthesis

TABLE 4

---

* The intermediate stages are:

$$2COOH \cdot CH_2 \cdot CHOH \cdot COOH \rightarrow 2COOH \cdot CH_2 \cdot CO \cdot COOH + 4[H]$$
$$2COOH \cdot CH_2 \cdot CO \cdot COOH \rightarrow 2CH_3 \cdot CO \cdot COOH + 2CO_2$$
$$2CH_3 \cdot CO \cdot COOH + 4[H] \rightarrow C_6H_{12}O_6$$

On addition,

$$2COOH \cdot CH_2 \cdot CHOH \cdot COOH \rightarrow C_6H_{12}O_6 + 2CO_2$$

i.e.
$$2C_4H_6O_5 \rightarrow C_6H_{12}O_6 + 2CO_2$$

This is the Wood-Werkman reaction. Alternatively the Ochoa reaction may be involved (page 122).

In interpreting any data involving R.Q.'s the limitations of the method must be remembered. In particular these are

(a) The possibility of two substrates being used simultaneously to give an intermediate value, e.g. if fats and carbohydrates are respired together, a value typical of proteins may be obtained.

(b) The effect of physical conditions, e.g. the different solubilities and diffusion coefficients of the two gases.

The Intensity of Respiration. This is usually measured in terms of either the CO_2 production or the O_2 consumption. In order that the value should be unambiguous the figure should also specify

(a) The temperature at which the measurements were made.

(b) The weight of the respiring tissue. This can be either fresh or dry weight.

(c) The time over which the measurements were made.

Thus a typical result would be x gm. CO_2/y gm dry weight/hr.

An alternative method depends on the loss of weight which occurs as a result of consumption of the respiratory substrate. In this case it is the dry weight which is important. The method is quite satisfactory for slices of storage tissue where the material is easily handled and the possibility of synthesis is excluded, but in general it presents more technical difficulties than those methods involving gaseous exchange.

The classical method of measuring the CO_2 production involves the use of a Pettenkoffer tube. A current of air is first passed through a soda lime tower (to remove CO_2) and then through a vessel containing the respiring material, which should be kept in the dark if there is the possibility of photosynthesis. The current of air then passes through the Pettenkoffer tube containing $N/100$ $Ba(OH)_2$. By back titrating with $N/100$ HCl, it is possible to calculate the quantity of CO_2 produced. If the rate of anaerobic respiration is required, it is a simple matter to pass nitrogen instead of air, in which case the soda lime tower is not required.

If it is desired to *compare* rates of respiration of different tissues, it is easier to measure the time taken for the CO_2 to effect a standard colour change of a suitable indicator such as Bromthymol blue.

External Factors Affecting the Rate of Respiration. The major external factors affecting the rate of respiration are

(a) Temperature.

(b) Oxygen concentration.

(c) Carbon dioxide concentration.

(d) Inorganic salts.

(e) Mechanical stimulation.

(f) Wounding.

Temperature effects upon the rate of respiration must be divided into initial effects and long-term effects. An excellent example of this is shown by Fernandes' data for four-day-old pea seedlings (see fig. 59). In his experiments the seedlings were initially all at a temperature of 25° C. They were then placed in their various experimental temperatures, ranging from 0° to 50° C and after three hours their respiratory intensities were measured. The final series of results showed that

(a) When the temperature was reduced, the rate of respiration decreased.

(b) When the temperature increased up to 45° C the rate of respiration increased.

(c) When the temperature increased above 45° C, the rate of respiration decreased.

In interpreting these results it will be noticed that the control group

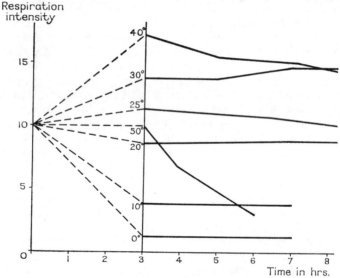

FIG. 59. Time–temperature effects on respiration intensity of four-day-old *Pisum sativum* seedlings. (From Stiles and Leach, after Fernandes)

of seedlings, maintained at 25° C, showed a slight increase in their rate of respiration for the first three hours and a slight decrease over the next four. This is a normal feature of seedling respiration and must be allowed for in considering the results at other temperatures.

From the first set of results it would be reasonable to conclude that over a temperature range of 0–45° C increase in temperature results in an increased rate of respiration and vice versa, while increasing the temperature above 45° C results in a fall in the respiratory rate. If the results are studied after a further four hours it will be seen that only at temperatures up to 30–35° C are the new respiratory rates maintained. At 40° C and above there is a definite falling off. These results can be attributed to the fact that prolonged exposure to higher temperatures causes damage to the cells' enzyme systems.

Measurements of Q_{10} are obviously complicated by this time effect and become meaningless at temperatures greater than 35° C. Within the range 0–35° C, the Q_{10} approximates to two, indicating temperature effects on enzyme controlled reactions.

Oxygen concentration. The function of oxygen is to reoxidize reduced coenzyme I and removing the necessity for the pyruvic acid–ethyl alcohol reactions and also to enable the Krebs' cycle to operate.

It would therefore be reasonable to expect that if the oxygen concentration of the surrounding air was lowered, then there would be a gradual reduction in the rate of respiration as measured by CO_2 output until eventually a stage was reached where some of the co-enzyme reoxidation was carried out by the pyruvic acid–alcohol system. This would mean that a mixture of anaerobic and aerobic respiration would result, the former would increase steadily and the value for the R.Q. would also increase. This explanation holds once

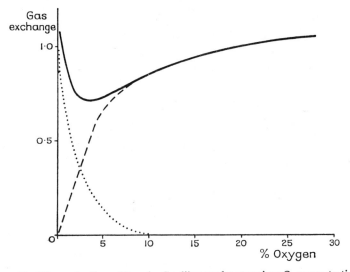

FIG. 60. CO_2 production of Bramley Seedling apples at various O_2 concentrations (rate in air = 1·0). Solid line = CO_2 emission; dotted line = anaerobic CO_2; broken line = oxygen consumption at low (<10%) concentrations. (From James, after Watson)

the cytochrome system ceases to be saturated with oxygen. At values above this level, the effect of oxygen on the mobilization of carbohydrate reserves may be important.

There is a considerable variation in the oxygen concentration at which different species are first affected but the data of Watson for Bramley Seedling apples (see fig. 60) is typical.

Beyond 10–15% O_2 concentration, where the solid line joins the broken line, O_2 consumption and CO_2 production follow the same course, i.e. R.Q. = 1. Increasing the oxygen concentration above that found in normal air results in a slight increase in the rate of respiration while decreasing the oxygen concentration produces a

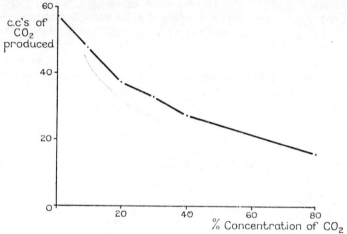

FIG. 61. Effect of CO_2 concentration on rate of respiration of germinating white mustard. (Based on data of Kidd, from Stiles and Leach)

slight decrease in the rate of respiration until a concentration of 10–15% O_2. At this point oxygen consumption continues to decrease steadily but the CO_2 production shows a sharp rise as anaerobic respiration starts. The progress of this anaerobic respiration is shown by the dotted line. Eventually, of course, when there is no oxygen present, all the respiration is anaerobic.

CO_2 concentration. CO_2 exerts a narcotic influence on plant cells and depresses respiration. Kidd in experiments with germinating mustard seeds found that increasing the CO_2 concentration to 80% (maintaining a constant 20% O_2) resulted in a marked fall in the rate of respiration as measured by the CO_2 output (see fig. 58).

The mode of action of CO_2 as a narcotic is imperfectly understood.

The presence of inorganic salts. The addition of salts to the external solution often increases the rate of respiration and the rate of increase is to some extent proportional to the rate of salt absorption

Lundegårdh attributes this increased respiration to the anions present and distinguishes two forms of respiration—a ground respiration and a variable anion respiration, the latter possibly being intimately connected with the cytochromes. This is discussed in greater detail under 'Salt Uptake' (page 78).

Mechanical stimulation. Audus found that if cherry laurel leaves were handled during the course of an experiment then there was an increase in the rate of respiration of from 20% to 183% which lasted for several days. If the treatment was repeated at intervals then there was a gradual diminution of the effect of the stimulus.

There is little doubt that this represents the effect of a purely mechanical stimulus, but so far no satisfactory explanations have been offered.

Wounding. If the surface of a plant is damaged there is an increase in the rate of respiration, quite independent of any localized increase in gaseous diffusion, for several days. It is interesting to speculate whether or not this increased respiration is associated with the metabolic activity involved in the action of secondary meristems and wound cork.

Photorespiration

The experimental work described previously assumes that in chlorophyllous tissue the rate of respiration in the light and dark are identical, i.e. that light does not have any effect on the rate of respiration.

It is obviously difficult to measure the rate of respiration in the light since photosynthesis will also be taking place. One method which has been used is to measure the uptake of $C^{14}O_2$ and $C^{12}O_2$ when the two are supplied together in equal concentrations to an illuminated leaf. If we make the assumption that the movement of each type of CO_2 is according to its own concentration gradient, then we would expect that the rate of absorption of the $C^{14}O_2$ will be greater than the uptake of $C^{12}O_2$ (see fig. 62) and the difference between the two values will provide a measure of the rate of respiration in the light.

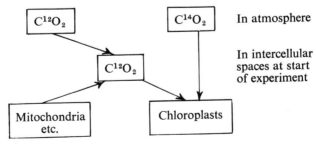

FIG. 62. Diffusion pathways in determination of photorespiration

It would appear that flowering plants can be separated into two categories, based on their respiratory response to illumination—

(a) Those which mainly utilize the Calvin cycle in photosynthesis: they may show appreciable photorespiration.

(b) Those which mainly fix their CO_2 by the dicarboxylic acid pathway: they show negligible photorespiration.

The respiration occurring in the light can come from three sources—mitochondria, pentose oxidation and peroxisomes (peroxisomes are cytoplasmic organelles containing some of the enzymes involved in glycollic acid metabolism).

The peroxisomal respiration is sensitive to light intensity (increasing with increasing light intensity) and involves the oxidation of glycollic acid ($CH_2OH\ COOH$). Subsequently the glyoxylic acid ($CHO\cdot COOH$) produced may form serine ($CH_2OH\ CHNH_2\ COOH$) from which sucrose can be formed.

The various stages involved are—

$$2CH_2OH.COOH + O_2 \longrightarrow 2CHO.COOH + 2H_2O$$
$$CHO.COOH + COOH(CH_2)_2CH.NH_2.COOH \longrightarrow CH_2.NH_2.COOH + COOH(CH_2)_2CO.COOH$$
$$CHO.COOH + [O] \longrightarrow H.COOH + CO_2$$
$$H.COOH + CH_2.NH_2.COOH \longrightarrow CH_2OH.CH.NH_2.COOH + [O]$$

On addition

$$2CH_2OH.COOH + O_2 +$$
$$COOH(CH_2)_2.CH.NH_2.COOH \longrightarrow CH_2OH.CH.NH_2.COOH + CO_2 + 2H_2O -$$
$$COOH(CH_2)_2.CO.COOH$$

The bulk of evidence favours the view that in plants—

(i) Mitochondrial respiration tends to be saturated at oxygen concentrations in excess of 2 per cent and is inhibited by relatively weak illumination.

(ii) Peroxisomal respiration continues to increase with increasing concentrations of oxygen (possibly up to 100 per cent O_2) and also increases with increasing light intensity.

9

Plant Hormones I

The Auxins

In the case of auxins, more perhaps than in other aspects of plant physiology, it is essential to consider the early work—to some extent it is true to say that we are no nearer an understanding of the mode of action of auxins now than we were thirty years ago. A great deal of this early work was concerned with an elucidation of phototropism and, to a lesser extent, geotropism.

Early Work on Phototropism. Tropic movements may be defined as bending movements of parts of a plant caused by differential growth resulting from the application of a unilateral stimulus.* The first observations were made by Charles Darwin (1880) and his experiments were confirmed and extended by Boysen-Jensen and Paal (1910–20). Briefly their experiments may be summarized as follows (see also fig. 63):

   (i) When a coleoptile was illuminated from one side only, the coleoptile grew towards that side.

  (ii) The bending took place some distance below the tip.

 (iii) Either removing the tip by amputation, or covering the tip by a small light-proof cap, resulted in the absence of a response following illumination. It was therefore concluded that—

    (a) The effect of light was perceived by the tip.

    (b) The stimulus was transmitted to lower regions where the bending took place.

 (iv) Whereas the ability to make a phototropic response was lost by amputation of the tip, it could be recovered if

    (a) The tip was replaced on the stump so that the two cut surfaces were in direct contact.

    (b) The tip was replaced but with the two cut surfaces separated by a thin piece of gelatin.

  (v) On the other hand, recovery of the phototropic response did NOT occur if cocoa butter, mica or platinum foil was used

* As opposed to *tactic* movements in which the whole organism moves in response to a unilateral stimulus (e.g. acellular algae) and *nastic* movements in which movements of part of a plant occur as a response to a diffuse or non-directional stimulus.

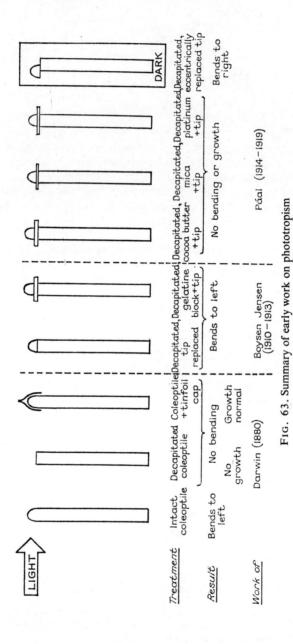

Fig. 63. Summary of early work on phototropism

instead of gelatin. From experiments (iv) and (v) Paal concluded that

(a) The stimulus was produced at the tip.

(b) The stimulus was water-soluble (if it had been fat-soluble it would have passed through cocoa butter and if it had been electrical it would have been transmitted across the platinum).

(vi) *In the dark* the eccentric replacement of the coleoptile tip resulted in a bending away from the side on which the tip was replaced. This suggested a possible explanation of the phototropic response, viz. in diffuse light the stimulus travels down the stem *in equal* concentrations on all sides causing an even stimulation of the growing region *but unilateral stimulation caused a greater concentration of the growth-stimulating substance on the unilluminated side, a greater growth rate on that side and a consequent curvature of the coleoptile.*

To explain the unequal distribution of the growth stimulant, Paal suggested three possible mechanisms:

(a) Inhibition of growth-stimulant production.
(b) Photochemical inactivation of the stimulant.
(c) Inhibition of downward movement on the illuminated side (and some lateral translocation).

The publication by F. W. Went of his '*Avena* test' in 1928 as a method of assaying the quantity of growth stimulant present in a given piece of tissue marked an enormous step forward. From the work of Paal and Boysen-Jensen it was obvious that the growth stimulant would pass from the decapitated tip into gelatin and then from the gelatin into the stem. Therefore, in order to isolate the stimulant Went placed the coleoptile tips on an agar (instead of gelatin) block standing on a piece of inert material such as glass. The stimulant passed into the agar and its presence could be demonstrated by cutting the block up into small cubes and placing them eccentrically on to decapitated coleoptile stumps kept in the dark. After a few hours a well-defined growth curvature was observed. An important development was that if the agar block was mixed with some stimulant-free agar so that there was a known dilution, then, over a wide range of dilution the *angle of bending was proportional to the concentration of the growth stimulant.* It was therefore possible to develop a *quantitative* approach to the subject and, with the isolation of the active compound, to calibrate the test.

In slightly modified forms the test is still in use, but other methods have been developed such as Went's split pea stem curvature test and Moewus' cress root growth-inhibition test.

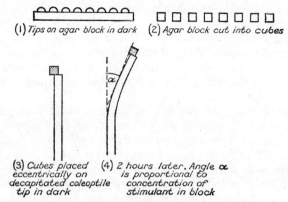

(1) *Tips on agar block in dark*   (2) *Agar block cut into cubes*

(3) *Cubes placed eccentrically on decapitated coleoptile tip in dark*   (4) *2 hours later. Angle α is proportional to concentration of stimulant in block*

FIG. 64. Essentials of Went's *Avena* coleoptile test

By an ingenious series of experiments using quantitative methods, Went was able to throw more light on the nature of the phototropic response. Coleoptiles were illuminated unilaterally and their tips *then* removed and placed on agar blocks separated into two halves by a safety-razor blade: a control series was set up consisting of tips which had not been illuminated. In this way it was possible to compare the production of auxins from the illuminated and non-illuminated sides of the tip (fig. 65).

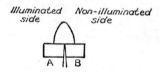

Illuminated side   *Non-illuminated side*

A   B

FIG. 65. Went's experiment with coleoptile tips

	Illuminated		Control (*dark*)	
	A	*B*	*A*	*B*
% Total growth stimulant	27	57	50	50

The amount of auxin entering A and B in the control was found to be equal in amount and the quantities found from the illuminated tips was expressed in terms of the amount from the non-illuminated. Two processes appear to be taking place:

(*a*) A *movement* of stimulant (7%) from the illuminated side.
(*b*) A *destruction* of stimulant (16%) on the illuminated side.

Went and his school attributed the phototropic effect mainly, if not entirely, to translocation rather than destruction, and in this they were supported by the results of van Overbeek who used pieces of hypocotyl sandwiched between two agar blocks, the upper containing a known quantity of stimulant and the latter being separated into two halves by a safety-razor blade. After exposure to unilateral

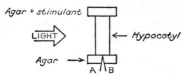

FIG. 66. van Overbeek's experiment

illumination it was found that the agar concentration in the lower block was greater on the side away from the light (B) than on the illuminated side (A) but there was *no evidence of any inactivation*. On the other hand Koningsberger and Verkaaik produced evidence to show that there was *no significant lateral transport*! Thirty years later there is still no unanimity of opinion as to the relative importance of the two methods: some of the evidence is discussed on pages 163–8.

The Hormone Concept. The early investigations of phototropism revealed the following features about its method of operation:

(*a*) The stimulus is transmitted as a chemical.
(*b*) It originates from a definite region (in this case the coleoptile tip).
(*c*) It is translocated.
(*d*) It causes a definite effect in a place remote from its site of production (viz. stimulation of elongation in the growing region).

These various points (a chemical stimulus—from a definite region—transported—producing a definite effect—in another part) are all characteristic of hormones. Further characteristics are:

(*a*) They are active in minute quantities.
(*b*) They can be extracted from the organism.
(*c*) The extract, on being supplied to the organism, can make good a deficiency caused by extirpation of the hormone-producing region.

The growth stimulants we have been considering satisfy all these criteria (formulated in the first place by animal physiologists) and so may be called *phytohormones* (plant hormones). They belong to the special type known as *auxins* (i.e. they are able to stimulate growth along the longitudinal axis of a stem depleted of its natural auxin supply). Other plant hormones are the gibberellins (which show some of the growth-stimulating properties of auxin) and the flowering hormones. (The latter are not, strictly speaking, entitled to the status of hormones since they have not yet been isolated.)

From the botanical point of view it is difficult to distinguish vitamins from phytohormones. Vitamins are manufactured by the plant, mainly in the leaves. All cells of the plant do not have this synthetic ability and they depend on a supply from other regions. They share with auxins all the characteristics of phytohormones even to the extent that, in tissue cultures, their absence results in typical abnormalities of growth which can be alleviated by the addition of the appropriate vitamin. In view of this the grounds for not considering them as hormones appear trivial—viz. in animal physiology they represent compounds which cannot be made by the animal's own metabolic machinery and in this respect they are quite distinct from hormones manufactured by definite endocrine glands. The status of vitamins as distinct from hormones has been retained by the plant physiologists although in this case the distinction is quite arbitrary and illogical.

Isolation and Chemical Structure of Auxins

In 1931 attempts were made by Kögl and Haagen Smit* to isolate material which showed activity in the *Avena* test. As starting points they used human urine and, in a later series of extractions, malt. Of the various substances they suspected of activity only one, which they called heteroauxin, has in fact been shown to be active and chemical analyses revealed that it was β indolylacetic acid (IAA).

β Indolylacetic acid

---

* The author is indebted to Professor M. H. van Raalte for drawing his attention to the results of mass spectrographic analyses of some of the original substances extracted. (Vliegenthart, J. A. and J. F. G., *Recueil des Travaux Chimiques des Pays-Bas*, **85**, 1966, pp. 1266-1272).

More modern methods using chromatographic techniques increasingly confirm IAA as the naturally occurring auxin and this is largely confirmed by molecular weight determinations obtained by measuring rates of diffusion through agar blocks. Many other compounds of similar structure have been tested for auxin activity (in the *Avena* test). Amongst those which were found to show significant activity were

α and β naphthylacetic acid (α NAA, β NAA).
Naphthoxyacetic acid.
various phenoxyacetic acids, especially those with chlorine atoms substituted in the benzene ring—2,4-dichlorophenoxyacetic acid, 2,4,5-trichlorophenoxyacetic acid and 2-methyl-4-chlorophenoxyacetic acid.

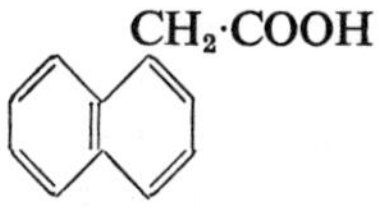

α Naphthylacetic
acid

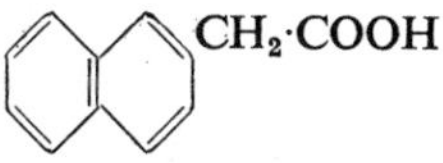

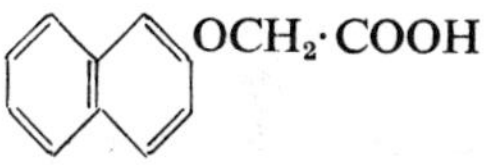

β Naphthylacetic
acid

Naphthoxyacetic
acid

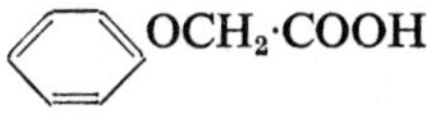

Phenoxyacetic
acid

2:4 Dichlorophenoxy-
acetic acid (2:4D)

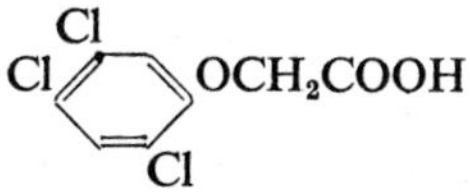

2:4:5 Trichlorophenoxy-
acetic acid (2:4:5T)

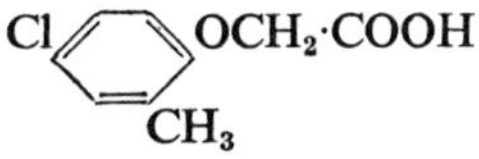

2-Methyl-4-chlorophenoxyacetic
acid

By studying these and similar compounds, attempts have been made to determine the essential structural requirements of a growth-stimulating compound—e.g. all the compounds shown share a double-bonded ring with a COOH group in the side chain which is separated from the ring by at least one C atom. From the analysis of large numbers of synthetic compounds many rules of structure have been proposed, all of them with a bewildering series of exceptions. One interesting feature of this type of work is that only two compounds have been reported to be growth stimulating and have a five-carbon ring—auxins *a* and *b*.

Implicit in the search for well-defined structural characteristics of growth-promoting compounds is the assumption that such sub-stances act directly (i.e. are not first changed into another compound) and that they all act on the same system. Since we do not know the locus of auxin action then obviously it is possible that this will explain the large number of exceptions.

Chromatographic extracts of plant tissues have revealed the existence of several other naturally occurring compounds with auxin properties including β indolylpyruvic acid, indole 3 carboxylic acid and β indolylglycollic acid, together with some unidentified non-indolyl compounds.

Auxin (IAA) Synthesis in the Plant

It is generally accepted that the IAA is synthesized at the growing points from a precursor which either originates from the storage organs of the seed (cotyledons or endosperm) or from the green leaves.

In support of this an experiment of Skoog can be cited. He placed an agar block on to the tip of a decapitated coleoptile still in contact with its endosperm. The agar block did not show any evidence of containing auxin when tested by the standard *Avena* method but, if the block was left on the test coleoptile stump, a delayed curvature resulted presumably due to auxin synthesis by the cells at the cut edge of the coleoptile from a precursor present in the agar block.

The most likely precursor is the amino acid tryptophan which is converted successively into β indolylpyruvic acid, β indolylacetal-dehyde and finally IAA.

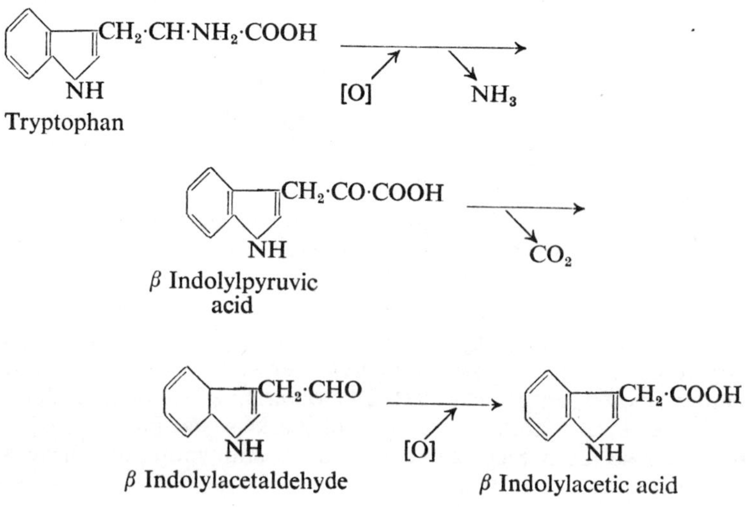

Tryptophan

β Indolylpyruvic acid

β Indolylacetaldehyde

β Indolylacetic acid

The evidence for such a reaction sequence is

(a) An enzyme system capable of catalysing the overall reaction tryptophan → β IAA has been shown to be very widely distributed in green plants.

(b) An enzyme system has been isolated and can be used to carry out an *in vitro* conversion.

(c) A low tryptophan content (e.g. in Zn-deficient plants) is associated with a low IAA content.

(d) β indolylacetaldehyde has been demonstrated in relatively high concentrations in plant tissues together with an enzyme capable of oxidizing it into IAA.

(e) β indolylpyruvic acid has been identified in chromatographic studies on coleoptile extracts and its conversion into IAA has been demonstrated in some micro-organisms.

Auxin Transport

Transport of auxin can be shown to take place both in the phloem and the parenchyma. Normally there is little lateral transport and it takes place in a morphologically downward direction, i.e. it is polar. This can be shown by Beyer's experiment in which a segment is cut from a coleoptile just below the tip. When the segment is replaced so that the morphologically upper surface is uppermost, normal growth takes place, but, if the segment is reversed, then there is very little growth.

A similar conclusion can be drawn from Went's experiment in which a section of coleoptile is placed between two agar blocks, the upper containing auxin (see fig. 67).

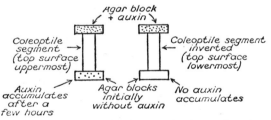

FIG. 67. Went's experiment on auxin translocation

From this type of experiment it is possible to show that the velocity of auxin transport is in the region of 10–15 mm per hour—a speed considerably faster than can be accounted for by diffusion but, on the other hand, appreciably slower than the velocity of sugar translocation. It can be demonstrated that aerobic conditions are necessary indicating a possible coupling of the process to respiratory energy.

Recent Work on Phototropism and Geotropism

The interpretation of the phototropic mechanism is complicated by the fact that the type of response varies with the intensity of the light stimulus used. DuBuy and Nuerenbergk were able to distinguish three positive curvatures and one negative curvature and the essential characteristics of these are summarized below.

Light intensity (metre candle sec)	Type of response*	Maximal sensitivity (in Avena)
Less than 4×10^3	*First positive* curvature	Apical 50μ
From 4×10^3 to 4×10^5	*First negative* curvature	Apical 3 mm
From 4×10^5 to 1×10^6	*Second positive* curvature	From apex to mid-
From 1×10^6 to 1×10^7	None	region
Greater than 1×10^7	*Third positive* curvature	Basal region

TABLE 5

The light intensity of normal daylight would normally be in the range causing third positive curvatures but most of the experimental work has been concerned with light intensities up to 1×10^6 metre candle sec. Unfortunately it is not always possible to tell from the published literature which particular curvature is under investigation

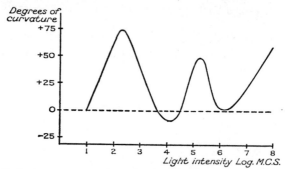

FIG. 68. Relationship of light intensity and phototropic curvature (after DuBuy and Nuerenbergk)

and a further complication is that the type of response to some extent varies with the part of the coleoptile illuminated.

Four main theories have been put forward to explain the role of auxin in the phototropic response: of these, Blaauws' light growth reaction can be dismissed since it basically invokes etiolation effects

* *Positive* curvatures are *towards* the stimulus or, when an agar block is placed eccentrically on a decapitated stump, towards the side on which the block is situated.

associated with the phytochrome system (see page 215). The three remaining theories are that phototropism results from—

(a) A light-induced *lateral translocation* of auxin.
(b) A light-induced *inactivation* of auxin.
(c) A light-induced *inhibition of auxin synthesis* at the apex.

This last theory, put forward by Galston, is a new idea. It suggests that as a result of such inhibition auxin precursor would accumulate on the illuminated side and diffuse to the unilluminated side. It has the great attraction that it could account for a reduction in the total quantity of auxin after unilateral illumination (b) and lateral translocation (a).

As most of the experimental work has been directed towards attempting to decide between the relative merits of the first two theories, only these will be discussed in detail.

Lateral Translocation as a Mechanism for Phototropism. In a series of measurements of the auxin concentrations at different light intensities, Wilden found that the highest auxin concentration was always on the side with the greatest radius of curvature (i.e. on the side furthest from the stimulus in a positive curvature and on the side nearest the stimulus in a negative curvature). Furthermore there was *no evidence for any destruction of auxin* (see table 6).

Curvature	% auxin on		Degrees of curvature
	Illuminated side	*Shaded side*	
1st positive	17	83	$+50°$
1st negative	62	38	$-15°$
2nd positive	36	64	$+20°$

TABLE 6

Boysen-Jensen developed an experimental technique in which a piece of platinum foil (later investigators have also used pieces of coverslip or mica) was inserted so as to divide the coleoptile tip longitudinally into two halves. He found that, for light intensities which would have induced the first positive curvature, curvature was inhibited if the foil was at 90° to the incident light but not if it was parallel to the incident light (fig. 69).

There is no reason why these results should occur if the curvature was due to photoinactivation; rather it indicates that the foil provides a barrier to lateral translocation.

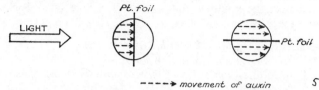

FIG. 69. Plan view to show position of Pt foil in Boysen-Jensen's experiment

When the illumination was of sufficient intensity to cause a second positive curvature bending took place. It is difficult to see how photoinactivation could account for the first positive curvature, but it is possible that it may be involved in the second positive.

This type of experiment has been extended in recent years by Winslow Briggs and his co-workers at Stanford University, U.S.A. They used, in addition to partly split tips, tips which were completely separated by thin glass barriers and they were able to show that for first and second positive curvature light intensities, unilateral illumination affected neither the total quantity of auxin nor its lateral distribution when the glass barrier was placed at 90° to the incident illumination. In the case of tips split only at their base it was

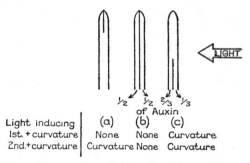

FIG. 70. Summary of split coleoptile experiments

found that the auxin concentration was about twice as great on the dark side than on the illuminated side, but there was *no evidence of any destruction*.

Another approach has been to supply the plant with IAA containing ^{14}C and then to measure the distribution of the applied auxin in terms of degree of radioactivity. Several experiments indicated that there was no transverse distribution of radioactivity across the coleoptile* as would be expected if there was lateral transport, but

* This result would give support to the photoinactivation theory since the number of C atoms in the inactivated compound would not have changed.

this is thought to be caused by fixation of auxin in the tissues. When the radioactivity of auxin collected in agar blocks from the basal ends of the coleoptile was measured, a difference in radioactivity was observed, being greater on the non-illuminated side (first positive curvature).

The case for lateral translocation would be further strengthened if a mechanism for sideways movement could be demonstrated. Since IAA is an acid it will show the property of *electrophoresis*, i.e. in an electric field it will move towards the anode.

Schrank was able to show a potential difference of about 15 mV at low light intensities (200 m c s) so that the illuminated side is negative with respect to the shaded side and a potential in the opposite directions for higher light intensities causing the negative curvature. Unfortunately when artificial potentials are applied across the coleoptile the behaviour of the auxin is not so consistent and in view of this Schrank has withdrawn his original suggestions.

Photoinactivation as a Mechanism for Phototropism. If photoinactivation is the cause of the lateral auxin gradient it should be possible to show

(*a*) A light-absorbing pigment with appropriate correlations of absorption spectra and action spectra.

(*b*) An *in vitro* photoinactivation system.

(*c*) An *in vivo* photoinactivation system.

If light is to provide energy for a photochemical reaction then it is essential that a pigment must be available both for the absorption of the light energy and its conversion into chemical energy. Carotene was the first pigment to be considered because

(*a*) It is present in the coleoptile.

(*b*) Its absorption spectrum corresponds closely with that of the action spectrum of phototropism (particularly for the absorption maxima at 4,500 Å and 4,800 Å and the minimum at 4,600 Å).

The other possible pigment is riboflavin which is also present in the coleoptile and has an absorption spectrum broadly similar to the action spectrum of phototropism. The case for riboflavin was strengthened by Galston's demonstration of increased light-growth-inhibition of etiolated pea stem segments when riboflavin was added to the IAA containing culture solution. He was also able to show, *in vitro*, that riboflavin was involved in the photo-oxidation of IAA, probably to indolylaldehyde.

$$\text{IAA} \quad CH_2 \cdot COOH + 2[O] \xrightarrow{\text{Photo-oxidation}} \text{CHO} + H_2O + CO_2$$

IAA β Indolylaldehyde

In an attempt to assess the relative importance of carotene and riboflavin, Reinert tried various combinations of IAA and the two pigments in *in vitro* photo-oxidations. His results (table 8) showed that, far from acting as the crucial light-absorbing pigment, carotene *inhibited* the effect of riboflavin almost completely, possibly because of the overlap of their absorption spectra. Confirmation of the unimportance of carotene comes from experiments with carotene-free

IAA	β Carotene	Riboflavin	Photoinactivation of IAA
Aqueous soln	Colloidal soln	Absent	None
Aqueous soln	Aqueous soln	Absent	None
Aqueous soln	Absent	Present	Photoinactivation
Aqueous soln	Aqueous soln	Present	None

TABLE 7

strains of albino barley which show normal phototropic responses.

These results, together with observations on the relative distribution of the two pigments in the coleoptile (fig. 71), led Reinert to develop his filter theory.

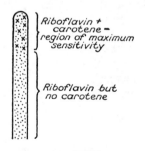

Riboflavin + carotene = region of maximum sensitivity

Riboflavin but no carotene

Riboflavin

Carotene

FIG. 71. Distribution of riboflavin and carotene in coleoptile

According to this theory the action of the carotene is to compete with riboflavin for light in the most effective wavelengths thus *decreasing the quantity of light and so of photoinactivation on the non-illuminated side.*

In support of the theory many experiments have been carried out using artificial filters. The coleoptile is a hollow cylindrical structure inside which can be found the rolled second leaf (fig. 72a).

Bunning used three groups of decapitated* coleoptiles (fig. 72b, c, d).

On exposure to unilateral illumination he found (b) showed a very weak positive curvature; (c) showed a stronger positive curvature

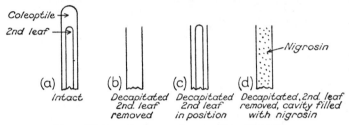

FIG. 72. Bunning's experiment with artificial filters

(about twice as strong as (b)); and (d) showed an even stronger response, approximately equal to that of the intact plant.

Although these experiments do not prove that carotene acts as an internal filter, it does suggest that such a filter operates.

The Relative Merits of the two Theories. The great weaknesses of the photoinactivation theory are

(a) The lack of evidence for an *in vivo* demonstration of photo-inactivation.

(b) The very large body of evidence which consistently fails to show any significant differences in the total quantity of auxin after unilateral illumination.

The weaknesses of the translocation theory are

(a) The experimental results (including the original experiments of Went) which show a *reduction* in the total quantity of auxin translocated during unilateral illumination.

(b) The lack of any evidence pointing to a *cause* for lateral translocation.

In a book at this level it is impossible to review all the work that has been carried out. The possibility undoubtedly exists that both mechanisms operate and that Galston's suggestion that light affects

* With only the tip removed (as opposed to the method used in the coleoptile test, where a second decapitation is made a few hours later) there will be some regeneration of auxin-synthesizing capacity so that weak phototropic curvatures can be obtained.

the synthesis of auxin from a precursor with a consequent modification of the translocation pattern may turn out to be correct. A final answer must also be capable of explaining phototropic reversals—for example, in ivy-leaved toadflax the flower stalk is positively phototropic, but after fertilization and with fruit development the stalk becomes negatively phototropic (forcing the fruit inwards towards the wall). Such a response might be a consequence of increased auxin production from the fruit.

The Phototropic Response of Leaves. Many leaves arrange themselves with the lamina at right angles to the incident light (are diaphototropic) and the mechanism, for the peltate leaf of *Tropaeolium*, has been elucidated by Brauner.

In this leaf the response is produced by differential growth of the petiole. By screening the leaf or the petiole he found (table 8) that the petiole was the sensitive region. Removal of the leaf blades caused the petiole to lose its phototropic response but this could be restored by applying IAA at a concentration of 1×10^{-6} to the cut ends. In this particular leaf it appeared that the lamina served as a source of IAA.

Leaf in	Light	Light	Dark
Petiole in	Light	Dark	Light
Phototropic response	Positive	None	Positive

TABLE 8

Geotropism

The ways in which the principal plant organs respond to gravity are shown in fig. 73. If, in addition to the responses shown on the diagram, it is remembered that

(a) Lateral roots and stems of 2nd order (i.e. branches off branches) or higher do not normally respond to gravity.

(b) The final position of the organs is controlled by the resultant of their geotropic response *and* their other tropic responses, viz.

(i) Most stems are positively orthophototropic (and most roots are aphototropic).
(ii) Many leaves are orthophototropic.
(iii) Most roots are positively hydrotropic.

then it will be seen that this ensures a plant in which

(i) The roots are well spread out and so are positionally adapted for anchorage.

(ii) The younger roots in particular will be in suitable positions for water absorption.

(iii) The aerial system is well spread out so as to provide a wide photosynthetic expanse.

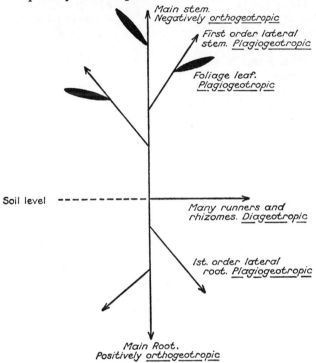

FIG. 73. The geotropic responses of the main vegetative organs

Whereas in the investigations of phototropisms it is a simple matter to vary the direction and the intensity of the light source, gravity cannot be altered so conveniently. It is possible to vary the effective intensity of the gravitational pull by placing the plant on a rapidly rotating wheel so that in addition to the gravitational force there is also an outwardly acting centrifugal force to which the plant also responds by a geotropic curvature. The final direction of growth will be along the resultant of the two forces. This was first carried out by Knight using a centrifugal force of about 3·5 g.

The principle of the clinostat is quite different. A vertical disc is rotated at a slow speed (c. 4 revs per hour) and seedlings are pinned on to the disc and kept moist either by being covered by small pieces of damp blotting paper or by covering the entire disc with a plastic cover containing water. Because of the rotation both upper and lower surfaces of the seedling are equally stimulated by gravity so that no geotropic response occurs. Confusion often occurs over the use of the word *control* in clinostat experiments. If by the term is meant the duplication of an experiment without the specific factor under investigation (so that the effect of the factor can be deduced by comparison with the main experiment in which the factor *is* present) then it follows that the *control* is the *rotating* clinostat.

Geotropism in Stems. When a stem is placed horizontally the negative geotropic response resulting in its upward growth curvature is generally attributed to an increased accumulation of auxin on its lower side. Several lines of evidence support this.

Occurrences associated with a high auxin concentration.

(*a*) Some pineapples can be induced to flower if the stems are kept in a horizontal position.

(*b*) Lateral bud development is normally suppressed on the lower surface of a horizontal stem.

(*c*) Adventitious roots are frequently formed on the lower surface of horizontal stems kept in contact with the soil.

That all these reactions can also be induced by the unilateral application of IAA indicates that there is an accumulation of auxin on the lower surface when the stem is placed horizontally.

Dolk's modification of Went's method for phototropism. Coleoptiles were placed horizontally for a length of time sufficient to produce a geotropic response. They were
then decapitated and placed *laterally* on agar
blocks divided horizontally by a razor blade (fig.
74). When the auxin concentrations in blocks U
and L were measured it was found that the concentration in L was markedly greater than in U.
Comparison with control experiments did not
indicate any auxin inactivation.

FIG. 74. Dolk's experiment

The logical explanation of the geotropic response of the stem is therefore a gravity-induced modification of the pattern of transloca-tion so that auxin accumulates on the lower surface of the stem with a consequent increased rate of growth in that region.

Particularly in the older work on geotropism there are many references to *statoliths*. Haberlandt suggested that the gravitational stimulus is perceived by the displacement of starch grains (statoliths) to the lower surface of the cell (the statocyte).

It is unlikely that the statoliths are of such a restricted nature as this and it has been suggested that other cell inclusions—nucleus, chloroplasts, mitochondria, or even large protein molecules—may function in this way.

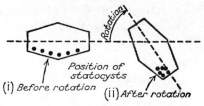

FIG. 75. Simplified diagram of statocysts, to show their position after a clockwise rotation of α°

There is evidence that the pattern of translocation becomes modified by the formation of electrical potentials (the underside of the stem becoming positive) so that the chain of reactions in geotropism might be

Gravity ⟶ displacement of statocysts ⟶ disturbance of cell membrane of lower surface ⟶ altered electrical potentials ⟶ modified translocation pattern ⟶ auxin accumulation on lower surface ⟶ negative geotropic curvature.

Geotropism in Roots

The gravitational response of roots is the opposite to that of stems. In this account the classical explanation is first described and then some of the more recent work is discussed.

The Classical Explanation. The experiments of Cholodny, Hawker, Moewus and others can be summarized as

(*a*) Maize roots continue to elongate after decapitation.

(*b*) Decapitated roots lose their ability to respond to gravity.

(*c*) Replacing the tip on a decapitated root results in a *retardation* of growth and restoration of sensitivity to gravity.

(*d*) Placing coleoptile tips on decapitated roots also results in a retardation of growth.

(*e*) Root tips placed on decapitated coleoptile stumps result in accelerated growth.

(*f*) Root tips placed eccentrically on coleoptile stumps cause positive curvatures.

(*g*) Agar blocks containing IAA placed centrally on decapitated roots cause a retardation of growth.

(*h*) Agar blocks with IAA, when placed eccentrically on decapitated roots, cause a negative curvature.

(*i*) Eccentric replacement of root tips on decapitated roots causes a negative curvature.

(*j*) Applications of *very low* concentrations (0·00001 to 0·001 μgm) of IAA to decapitated roots leads to an acceleration of growth.

(*k*) When Dolk's method is applied to horizontally placed roots, the concentration of auxin is greatest on the lower surface (cf. stems).

From these results it can be concluded that

(*a*) IAA is the auxin of both roots and shoots (experiments (*d*), (*e*), (*f*)).

(*b*) Increasing the IAA level, or even bringing it up to its normal value, causes an *inhibition* of root growth (experiments (*c*), (*g*), (*h*) and (*i*)).

(*c*) Very low concentrations of IAA accelerate root growth (experiment (*j*)).

(*d*) Gravity causes a modification of the translocation pattern in roots, as in stems (*k*), so that the auxin accumulates on the lower surface, inhibiting growth there with a bending taking place in a downward direction.

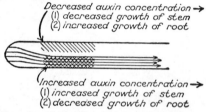

FIG. 76. Translocation path of auxin in horizontal root or stem

The different sensitivities of roots, buds and stems are summarised in fig. 77 from which it can be seen that—

(*a*) Applied auxin concentrations which stimulate root growth have no effect on stem growth.

(*b*) Concentrations which stimulate stem growth have an inhibitory effect on root growth.

(*c*) Even higher concentrations will also inhibit stem growth.

Recent Work on Geotropism. There are several pieces of work which throw doubt on the validity of the classical theory although it is not yet possible to offer an alternative explanation. The work is described under four headings:

The validity of root decapitation experiments was investigated by Younis. In an extensive series of decapitation experiments with broad bean seedlings, he

(*a*) Was not able to observe an accelerated rate of growth following decapitation.

(*b*) Confirmed that there was a reduction in sensitivity to gravity.

(c) Did not observe that replacement of the tip restored sensitivity to gravity.

Effects of anti-auxins* on root growth. If the roots' response to auxin is caused by the auxin being present in excessive quantities, then it might be expected that the addition of anti-auxins, since they

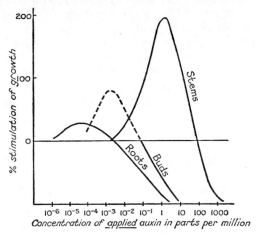

FIG. 77. Effect of auxin on growth rates of root stem and buds (after Audus)

are antagonistic to auxin action, would serve to reduce the effective auxin concentration and so promote root growth. In many cases this does not happen, but instead the anti-auxin shows the same kind of effect on root growth as does IAA.

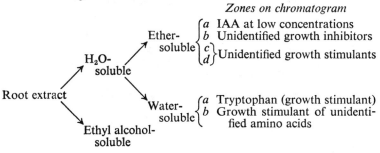

FIG. 78. Outline of extractions for chromatographic analysis

* Anti-auxins are compounds which compete with auxins for the reaction centres of the cell in the same way as enzyme inhibitors compete with enzymes for positions on the substrate surface. Thus they have a chemical structure very similar to, but not identical with, that of auxins. Obviously such compounds will greatly modify the action of auxins in the plant.

Chromatographic analysis of root tissues. This approach was initiated by Bennet Clark. The extract was divided into water-soluble and alcohol-soluble fractions (fig. 78) and the former was further divided on the basis of ether solubility. Separation of the various constituents of each fraction was carried out by paper chromatography and the constituents were extracted from the chromatrogram and tested for auxin activity in terms of their ability to stimulate the extension of coleoptile segments. Little activity was found in the alcohol-soluble fraction, but the important feature is that at least six types of auxin compound are present in the root—IAA, a growth inhibitor, two unidentified (ether-soluble) growth stimulants, tryptophan, and several unidentified amino acids. It seems, from this work, that the control of root growth is unlikely to be a single hormone mechanism.

Tissue cultures of roots suggest that root growth is controlled by a *hormone balance* in which IAA, the gibberellins and kinetin are involved. This is discussed on page 218.

Other Effects of Auxins

Although the definition of auxins stresses their importance in stimulating the elongation of a stem, there are many other effects caused by auxins, some of which are of great practical importance in agriculture and horticulture. A simple classification of auxin effects, apart from those already described, is shown below.

MAIN EFFECT	CONSEQUENCES
AUXIN { Stimulation of cell division (mitosis and cytokinesis)	Stimulation of cambial activity / Formation of wound tissue / Root initiation / Stimulation of fruit growth / Parthenocarpic fruit development
Maintenance of cell wall integrity	Inhibition of leaf abscission / Inhibition of fruit abscission
Inhibition by high concentrations	Inhibition of lateral bud development / General growth inhibition—hormone weed killers

The Stimulation of Cambial Activity by IAA. Evidence for this includes

(*a*) The seasonal activity of the cambium in a plant is closely paralleled by a similar variation in auxin synthesis by the developing buds.

(*b*) In decapitated seedlings, cambial activity ceases but it can be restored by the application of an agar block containing IAA.

Although the effect of IAA stimulation on cambial activity is well established, the actual *differentiation* of xylem and phloem needs the presence of gibberellins (page 195).

Another feature of cambial activity is the production of 'wound tissue' (callus) if it becomes exposed by damage to the overlying bark. This is a defence mechanism preventing excessive water loss and the entry of pathogens, but it can also be put to practical use in grafting, where the callus plays an important part in strengthening the union between stock and scion. It is difficult to assess the direct effect of the auxin on the formation of wound callus—its ease of formation is linked with cambial activity but on the other hand the damaged tissue almost certainly releases a chemical (e.g. traumatic acid, $COOH \cdot CH : CH \cdot (CH_2)_8 \cdot COOH$) which stimulates callus formation by the cambium. In the case of many grafts the ease of union and its strength is stimulated by IAA—except when callus formation is so enhanced that it actually pushes the scion away from the stock.

Root formation and IAA concentration has already been mentioned on page 172 in connexion with the greater amount of auxin found on the lower surface of a horizontal stem. Application of IAA to a root, apart from its effect in reducing the growth, causes an abundant production of laterals and this can be correlated with stimulated cell division in the pericycle. The formation of *adventitious* roots can be enhanced by similar treatment of the stem, but in this case the cells forming the root initials have already differentiated into 'permanent' tissue (as opposed to the root pericycle cells which are essentially undifferentiated promeristem). Under natural conditions the auxin supply for the formation of adventitious roots originates either from the leaves or the expanding buds.

The importance of hormones in promoting rootings for horticulture often depends not so much on the number of roots formed as on the speed with which they are developed. For commercial use naphthylacetic acid and γ indolylbutyric acid are markedly superior to IAA (fig. 79).

Stimulation of Fruit Growth and Parthenocarpic Fruit Development. *True* fruits originate from the ovary wall and *false* fruits from other tissues (e.g. the receptacle) by extensive growth after fertilization. At first the growth is mainly by cell division but in later stages cell enlargement predominates.

Since fertilization normally stimulates fruit development it is an obvious conclusion that the pollen grain contains either auxin or a substance which reacts with something present in the ovule to form auxin. An auxin can be extracted from pollen grains but it is not found in sufficient quantities to account for either the necessary amount of growth or the large rise in auxin activity detected after fertilization. In some cases (e.g. rye, apple) it can be shown that this auxin is produced by the rapidly developing endosperm. In these cases the function of the auxin in the pollen is to stimulate endosperm development with the consequent formation both of a food store and a site for auxin synthesis.

In natural parthenocarpic fruits (fruits formed without fertilization such as some varieties of banana) the auxin supply probably originates directly from the ovule; in the banana the natural seedless varieties usually have rather larger ovules than those with seeds and this might well offer an explanation of the greater auxin production.

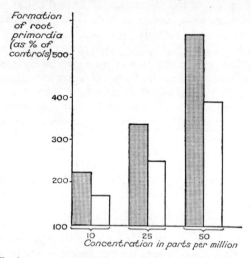

FIG. 79. Effectiveness of naphthylacetic acid (shaded) and IAA (plain) in formation of root primordia in bean seedling. (Based on data of L. C. Luckwill, *J. Hort. Sci.*, 1956, **31**, 89–98)

By the artificial application of auxins, in the form of sprays and aerosols, it is possible to raise the auxin level sufficiently to induce parthenocarpic development (e.g. tomatoes, apples, pears, strawberries, blackberries). The benefit of such treatment is mainly in augmenting the effect of normal pollination—the resulting fruit is not always seedless and in some cases flavour is impaired.

Inhibition of Abscission. The phenomenon of abscission applies to both leaf and fruit fall and in both cases is produced by the development of a *separation layer* in the abscission zone at the base of the petiole, pedicel or peduncle. In the separation layer there is a weakening of the middle lamella so that the cells are easily separated from each other by quite small mechanical disturbances.

The effect of auxin is to prevent the development of the separation zone and is presumably by a similar sort of action to those described on page 180 under the effects of auxin on the wall.

If the lamina is removed from the petiole a separation zone will develop quite quickly, but this can be inhibited by the application of IAA or NAA in lanolin paste at the distal end of the petiole, suggesting that under normal conditions the blade produces a supply of auxin which maintains the integrity of the abscission zone; with increasing age of the leaf the auxin supply decreases and so the separation zone develops.

The auxin supply from the fruit will similarly inhibit abscission. There are two stages when the auxin level may be low. The first is just after fertilization and the second is immediately prior to full fruit development. At either of these times abscission is possible with disastrous economic consequences for the fruit grower. Spraying with auxin (IAA, NAA or 2:4D) can raise the level sufficiently to inhibit abscission.

Inhibition of Lateral Bud Development. With bud development, into vegetative or floral shoots, it is usual to find that whereas the apical bud develops, the lateral (axillary) buds remain dormant. If however the apical bud is removed then the laterals will develop, i.e. the plant shows *apical dominance*. When IAA in lanolin paste is applied to the cut surface from where the apical bud was removed, then the laterals do not develop: in other words the action of the apical bud can be replaced by a suitable supply of auxin. The problem which remains is to explain why the lateral buds should be inhibited at a level of auxin associated with apical development. It is possible that they might have different levels of sensitivity (cf. roots and stems) but it is also likely that nutrition plays a part—in flax grown with a high nitrogen supply the application of auxin to the apex has very little effect on lateral inhibition whereas under poor nutritive conditions, apical dominance is well developed.

Hormone Weed Killers. The extreme sensitivity of the root to auxin had been known for a long time before plant physiologists considered the possibility of using such compounds as herbicides. The first research was done in the early days of the Second World War in an attempt to increase food production. Among the most effective compounds are

2:4D   (2,4 dichlorophenoxyacetic acid)
2:4:5T   (2,4,5 trichlorophenoxyacetic acid)
MCPA   (2 methyl 4 chlorophenoxyacetic acid)

The concentrations used are about a hundred times greater than normal physiological concentrations and under these conditions their effect is to overstimulate the growth-promoting activities of the cell to such an extent that they are completely thrown out of balance. Localized rapid areas of growth in the roots result in them becoming grossly distorted with a consequent easy entrance for pathogens and in some cases the blockage of sieve tubes by local pressure. Accompanying these structural abnormalities there may be derangements of mitosis with the production of unviable cells with unbalanced chromosome numbers.

Because of their auxin properties these compounds are easily translocated so that their effect is observed over the whole plant. For the most part they are absorbed more readily by dicotyledons than by monocotyledons and this makes them particularly useful as weed killers when applied to cereal crops.

Special precautions must be taken when they are used as herbicides for dicotyledonous crops since both crop plant and weed are susceptible. In the case of sugar beet the seeds can be sown with an adsorbent (e.g. activated charcoal) and the soil is then sprayed with the herbicide. The hormone is *ad*sorbed in the vicinity of the seeds but elsewhere it is available to be *ab*sorbed by the weeds.

Unfortunately it is becoming increasingly clear that such methods of weed control are fraught with great hazards. After all, a weed is only 'a plant growing in the wrong place' and the indiscriminate use of this kind of chemical can cause untold ecological havoc. Restricting examples to those with the possibility of economic consequences, some of the possible ways are

(*a*) By a toxic effect on insects and domesticated animals either by eating the chemical or by eating a plant whose metabolism has been altered.

(*b*) By the eradication of the food plants of insects.

(*c*) By the recolonization of cleared areas by herbicide-resistant weeds.

(*d*) By the evolution of herbicide-resistant varieties of weeds.

The last two examples parallel the processes by which antibiotic-resistant strains of pathogenic micro-organisms have developed. The emphasis on insect survival is made because of the importance of these animals for cross-pollination.

The Mode of Action of Auxins

In this section only the action of auxin in promoting cell elongation will be considered, and the term auxin will be taken to mean IAA. No mention will be made of theories which are based on studies of the structure of auxin molecules.

During the elongation of a cell the most obvious change is in the development of the cell vacuole and the increase in volume of the cell sap. Water enters the cell because of a diffusion pressure deficit. DPD, which can be represented as

$$DPD = (OP_i - OP_e) + WP + A$$

(using the abbreviations on page 43).

The process of vacuolation forms an integral part of cell elongation; since the latter is stimulated by IAA it is not unreasonable to expect that water uptake, which can be achieved by increasing the DPD, will be similarly enhanced. There are three ways this could be done:

(a) By an increase in the value of OP_i.
(b) By a decrease in the value of WP.
(c) By an increase in the value of A.

Evidence for an Auxin-induced Increase of OP_i. Measurements of the osmotic activities of elongating cells show that, on the whole, there is a decrease in the osmotic pressure of the cell sap, but the evidence is scanty. Usually the *quantity* (as opposed to the concentration) of osmotically active materials remains constant, although in some cases there is a slight increase. There is no evidence to suggest that any increase in osmotically active substances is ever adequate to account for the necessary influx of water.

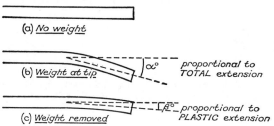

FIG. 80a. Principle of Heyn's determination of elasticity and plasticity

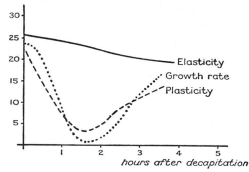

FIG. 80b. Changes in growth rate, elasticity and plasticity of an *Avena* coleotype after decapitation (Heyn, 1932)

Evidence for an Auxin on Cell Walls. The pioneer work in this field was done by Heyn in his measurements of the plastic and elastic extensibility of *Avena* cell walls. The principle is shown in fig. 80*a*. A horizontal decapitated coleoptile is bent by placing a weight at the end. The angle of bending, $\alpha°$, is measured. The weight is then removed and the new angle of bending is measured.

The *total* extensibility is proportional to the angle $\alpha°$. Of this, some, $\beta°$, is irreversible (plastic extensibility) and some reversible, $(\alpha - \beta)°$; this is known as the elastic extensibility. By applying IAA to the decapitated tips it was possible to measure growth rate, plasticity and elasticity. The results are shown in fig. 80*b*, and it will be seen that there is a very close correlation between growth rate and plastic extension but very little correlation between growth rate and elastic extension. Thus the effect of auxin on growth was visualized as a lowering of the wall pressure with a consequent increase in DPD.

The results of Heyn's work should be considered in conjunction with that of Bonner on the dry weights of coleoptile *walls* (table 9).

Treatment	Final length in auxin as % of control	Final wt in auxin as % of control
25° C	115	116
2° C	110	101
1% fructose at 25° C	128	142

TABLE 9. Data of Bonner (*Proc. Nat. Acad. Sci.*, 1934, **20**, 393–7) on the effect of temperature on the growth of coleoptile walls

The main conclusions which can be drawn from Bonner's work are

(*a*) At 25° C (without fructose added) the rate of elongation parallels the rate of synthesis of new wall materials.

(*b*) At 25° C (with added fructose) the formation of cell wall material exceeds the rate of elongation.

(*c*) At 2° C the *elongation takes place with negligible formation of new wall material.*

Or, from a simpler point of view, at 2° C the wall would become thinner, at 25° C without added fructose it would remain at constant thickness, but when fructose is added it would increase in thickness. It follows from this that growth must take place in at least two stages: a stretching in which there is no new material formed, followed by the production of new wall substance. It is the first process which is affected by auxin.

Broadly similar views have been deduced by Burstrom from his experiments on root growth. He thinks that the second stage, involving synthesis, depends on the presence of Ca^{2+} ions and may be inhibited by auxin.

If the wall itself is examined it is found to become thinner in the early stages of growth (as would be expected if extension is taking place) followed by a stage where the thickness increases to a greater value than the original (i.e. when new materials are added). During growth there is a uniform extension over the whole length of the cell and the probable mechanism is first a loosening of the connexions between the cellulose micellae and the subsequent insertion of new materials, i.e. growth by *intussusception*.

Lee and his co-workers have recently (1967) isolated cell walls from the adjacent cytoplasm and they found that about one third of the material was subjected to spontaneous autolysis, thus showing that enzymes were present in the actual wall material. The enzymes are thought to hydrolyse the hemicelluloses present in the wall. Since such enzymes (glucanases) have also been shown to accelerate the growth of oat coleoptiles, it is tempting to speculate that auxins may exert their effect on wall plasticity by stimulating enzyme production or action.

Auxin Effects on Active Water Uptake

Reference to active (non-osmotic) water uptake has already been made. The first work was done by Reinders in 1938 with potato tuber discs. He found that if, after 24 hours in stagnant tap water, he transferred them to aerated distilled water, they showed a considerable increase in fresh weight. The addition of IAA to the distilled water resulted in an even greater increase in fresh weight and an *enhanced decrease in dry weight*. Reinders equated the enhanced decrease in dry weight with an increased rate of respiration and suggested the reaction sequence

IAA $\longrightarrow$ stimulation of respiration $\longrightarrow$ stimulated active H_2O uptake

Commoner and his co-workers also produced results which indicated the possibility of water absorption against an osmotic gradient although they produced an explanation linking water absorption with salt uptake—which would not explain Reinders' results. In their experiments they found that the loss of fresh weight caused by placing potato discs in hypertonic (0·2M) sucrose could be offset by the addition of IAA and the further addition of either potassium chloride or potassium fumarate actually caused an *increase* in fresh weight.

Repetition of these experiments under aseptic conditions by van Overbeek confirmed the auxin effect on water absorption but did not find any additional water uptake when salts were added. This of course agrees with Reinders' work.

The Effects of Auxin on Respiration. If auxins are concerned with water absorption, either by modifying wall structure (and there is *no* evidence to suggest that they in any way combine with parts of the wall) or by causing a non-osmotic water uptake, then they probably exert this effect by a stimulation of respiration. A great deal of experimental work on this has been done by Bonner and some of the more important conclusions are

(*a*) *Avena* coleoptiles do not grow under anaerobic conditions.

(*b*) Cyanide inhibits coleoptile growth.

(*c*) The addition of IAA often leads to increased oxygen consumption.

(*d*) Dinitrophenol (DNP) inhibits growth by 90%, increases oxygen consumption and also inhibits auxin-induced water uptake.

Before concluding from this that the sequence of events is substantially the same as that suggested by Reinders it is necessary to examine the mode of action of DNP. This is to 'uncouple' oxidative phosphorylation in the cell from electron transfer via the cytochrome system.

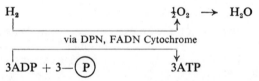

$$H_2 \qquad\qquad\qquad \tfrac{1}{2}O_2 \;\rightarrow\; H_2O$$

via DPN, FADN Cytochrome

$$3ADP + 3-\text{\textcircled{P}} \qquad\qquad 3ATP$$

Since these two reactions are linked, a *lack of phosphate acceptor* (ADP) can inhibit the rate of hydrogen transfer and therefore limit the rate of respiration. When DNP separates the two reactions (a simple analogy is the depression of the clutch pedal in a car separating the engine from the transmission) it would follow that the addition of DNP could

(*a*) Accelerate the rate of respiration by freeing it from its dependency on ADP.

(*b*) Decrease the availability of $\sim$⒫ groups and so bring endergonic reactions to a halt.

In view of the mode of action of DNP the most likely way in which IAA affects respiration is indirectly by *stimulating the utilization of* $\sim$⒫ and so producing a higher level of ADP.

Other Possible Auxin Effects. There is quite substantial evidence that auxin causes a *decreased protoplasmic viscosity* which is paralleled to some extent by an *increase in the rate of protoplasmic* *streaming.* Attempts have also been made to try and detect *increases in permeability of protoplasmic membranes*; on the whole this has not been very successful although it seems likely that permeability to amino acids is increased.

If these effects are important they would presumably act by accelerating rates of translocation and so increasing the availability of respiratory substrates. It is interesting to note that a postulated mode of action of the animal hormone thyroxin is by increasing the permeability of the mitochondrial membrane.

Fig. 81 represents a tentative attempt to show some of the possible modes of auxin action and how inter-relations between processes could produce some of the observed results.

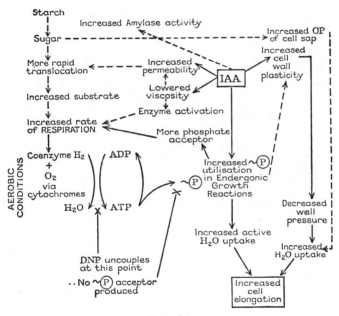

FIG. 81

10

Plant Hormones II:
Florigen and Gibberellins

Florigen

It must be emphasized at the outset that the flowering hormone, florigen, unlike the auxins, has not been isolated and therefore is not, in the strictest sense of the word, to be considered a hormone although the evidence for its existence is very convincing.

The Effect of Daylength on Flowering. Early ideas of the causes of flowering were mainly concerned with the effects of nutrition. A particularly good example of this can be found in the work of Kraus and Kraybill who found, in the case of tomato plants, that a good supply of nitrates and carbohydrates (i.e. good conditions for photosynthesis) favoured vegetative growth, reduction of nitrates caused a swing towards reproduction, but with a poor nitrate supply both reproduction and vegetative growth declined. Although this work has led to useful generalizations for the horticulturist it has not opened up any useful avenues of exploration for the physiologist.

The first evidence for a *daylength* effect was produced by Garner and Allard in 1920. They were experimenting with genetical crosses of tobacco plants and to do this they needed varieties which flowered at the same time. The variety Maryland Mammoth would not flower in summer but if the plant was kept in a greenhouse (so that vegetative growth continued and there was protection from the cold autumn nights) it would flower in October or November. In what seems to have been a last desperate effort to get the plant to flower in summer, the light supply to half of them was reduced to seven hours a day from early July. These flowered in late August but the control plants did not flower until October. Furthermore, seeds sown in November produced plants which flowered in April. The obvious explanation was that *flowering was caused by exposure to days made up of short light and long dark periods.*

Extension of this work showed that it was possible to divide plants into three categories—'short day', 'long day' and 'day neutral'.

Short day plants were originally defined as plants which would only flower if the daylength was less than twelve hours. This has been shown to be an over-simplification and it is better to consider them as plants in which the *daylength must not exceed a certain critical value* if they are to flower.

Long day plants on the other hand were defined as plants which would only flower if the daylength was greater than twelve hours but it is better to consider them as plants which will only flower *if the daylength exceeds a certain critical value*. They will flower in continuous light and periods of darkness have an inhibitory effect.

Day neutral plants are plants in which flowering is unaffected by daylength.

Some of the plants in these categories are shown in table 11. It includes some common plants and plants used in research.

Short day	Long day	Day neutral
Chrysanthemum *Xanthium* (cocklebur) Soybean Strawberry Spring ⎫ flowering plants* Autumn ⎭	*Hyoscyamus* (henbane) Beet Radish Wheat Summer flowering plants*	Tomato Cucumber Maize Dandelion Tropical plants†

TABLE 10

* Refers to plants of temperate climates.
† Refers to many plants which flower all the year round.

Short Day Plants

Light–Dark Cycles. The number of light–dark cycles varies between plants but there is no need for the treatment to be continued until flowering. In the case of cocklebur it has been found that *one* cycle will induce flowering (i.e. after one cycle the plant can be kept in continuous light and flowering will take place) but as the number of cycles is increased, the time taken for flowering will decrease.

Daylength. Increasing the length of the light period in the cycles, even although there is an adequate dark period, does not necessarily increase the amount of flowering or decrease the time for flowering. In experiments with soybean Hamner found that with a constant dark period of sixteen hours, flowering would occur in continuous light after seven light–dark cycles providing the light periods exceeded five hours. Maximum flowering was achieved when the light periods were from ten to twelve hours, but there was no flowering if the light was extended to twenty hours per cycle.

During the length of time when the light has an acceleratory effect on flowering the effect of the light is proportional to its intensity. The inhibitory effect can be brought about by very low light intensities.

Carbon Dioxide. CO_2 is necessary during the light stage if it is to be effective. It can be shown that there is no flowering in CO_2-free air, and the amount of flowering can be increased if the concentration of CO_2 is raised above that of air. Details of the fate of this CO_2 are obscure. The majority opinion favours the view that it is not associated with photosynthesis since there seems to be little direct correlation between flowering and the carbohydrate level of the plant. On the other hand there is evidence to suggest that the CO_2 may be dispensed with if the plant is supplied with additional sugars or Krebs' cycle acids. Some CO_2 is absorbed during the dark stage but its significance to flower formation is not known.

Temperature. The effect of low temperature is to inhibit flowering. The effect is greatest if the low temperature occurs in the dark stage. Since the Q_{10} of both light and dark stages exceeds unity this cannot be explained by photochemical reactions being insensitive to temperature. It is usually considered as evidence that the majority of the synthetic reactions occur in the dark.

Interruption of Cycle. The effect of interrupting the dark period by light is drastically to inhibit flowering. The most significant features associated with this are that:

(a) It is *least* effective if given near the beginning or the end of the dark period.

(b) It is *most* effective if given at a time which corresponds to the critical dark period.

(c) Only a low light intensity is necessary, although the effect is greater at high light intensities.

(d) Red light is most effective (with maximum effect at 6,500 Å).

(e) Infra-red light is least effective (with minimum effect at 7,350 Å).

Perception of Light Stimulus. The perception of the light stimulus occurs in the leaves. This was first demonstrated by Cajlachjan working with *Chrysanthemum* and *Perilla*. In his experiments with *Chrysanthemum* he grew the plants under long day conditions and removed the apical bud so that there was a profuse development of potentially flower-bearing lateral shoots. The laterals were defoliated so that only the lower part of the stem bore leaves. He divided his plants into four groups and, by the use of light-proof cases, each group received a different regime of light–dark cycles.

Group A. Entire plant continued to receive long day treatment.
Group B. Lower leafy region received short days: upper defoliated region received long days.

Group C. Lower leafy region received long days: upper defoliated region received short days.

Group D. Entire plant received short day treatment.

He found that in plants in which the *leaves* received short day treatment, flowering occurred (B and D), but when the leaves received long days (A and C), there was no flowering. It would appear that

(*a*) The short day stimulus is perceived by the leaves.

(*b*) The stimulus is transmitted to the buds. Cajlachjan called this stimulus '*florigen*'.

In a further series of experiments with *Perilla* plants grown in long days, he used a plant that had been defoliated except for one leaf. By means of differential shading he was able to give *either* short days *or* long days *or* continuous darkness to *either* the whole leaf *or* to the proximal half of the leaf *or* to the distal half of the leaf or to a longitudinal half, one side or the other, of the midrib.

The results are summarized in table 11.

Distal/Proximal shading	Light treatment								
	1	2	3	4	5	6	7	8	9
Distal half	S	S	S	L	L	L	D	D	D
Proximal half	S	L	D	S	L	D	S	L	D
Flowering ?	√	×	√	√	×	×	√	×	×

Longitudinal shading	Light treatment					
	10	11	12	13	14	
Left half	L	L	L	S	S	D
Right half	L	S	D	D	S	D
Flowering ?	×	×	×	√	√	×

TABLE 11. Cajlachjan's experiment with *Perilla*
(S = short days, L = long days, D = darkness, √ = flowering
× = no flowering)

This not only confirms that the leaf is the receptive organ—it also introduces a new conception, viz. if there is a portion of the leaf in long day conditions either between the short day region and the petiole or longitudinally adjacent to the short day conditions (condition 11), then the effect of the short day stimulation is lost. When this is considered in conjunction with experiments showing the effect of interrupting the dark period, it appears that the product of the short day conditions is sensitive to excess illumination whilst still in the leaf.

There is evidence to show that the *age of the leaf* is important. Very young or very old leaves are not sensitive to photoperiodic treatment.

A period of high light intensity is necessary after the dark period (in addition to the light period already discussed which precedes the dark period). The evidence for this has been obtained from cocklebur where it has been shown that

(*a*) The number of flowers formed when a twelve-hour dark period is followed by a light flash and a further five hours dark is less than when the twelve-hour dark period is followed by five hours light and then five hours dark (i.e. when there are five hours light instead of a light flash).

(*b*) No flowering results if the leaves are cut off after the 12 hours dark + 5 hours light which indicates that at this stage the stimulus is still in the leaf.

(*c*) The effect is proportional to the light intensity.

(*d*) A supply of sucrose can be substituted for the light.

The *modus operandi* of this second high light intensity reaction is thought to be *either* a stabilizing effect on the stimulus produced in the dark *or* a means of 'boosting' the stimulus out of the leaf by increasing the rate of mass flow (although, as we have seen in chapter 4, there is no overwhelming evidence for a mass flow hypothesis).

Hormones. The evidence that a hormone is involved comes from several sources, some of which have already been mentioned:

(i) The spatial separation of the site of stimulation and the position of the response. The stimulus is perceived in the leaves, although the actual response, by which a bud changes into a flower, occurs at a distance.

(ii) Defoliation after photoinduction may or may not result in flowering. In general it is found that if the leaves are removed very soon after induction there is no flower production whereas there is if the leaves are removed after an interval of several hours. Presumably the stimulus does not pass out of the leaf as soon as it is formed —defoliation before translocation inhibits flowering.

(iii) The results of grafting experiments are shown in a summarized form in fig. 82. As long as a graft is possible then a flower kept in non-inducting conditions (e.g. a short day plant kept in long days or a long day plant kept in short days) can be induced to flower if a photoinduced plant is grafted on to it. All grades of grafting are possible—intravarietal, intervarietal, interspecific and intergeneric. Grafting short day plants on to long day plants and vice versa are equally successful. Not only does this support the idea of a hormone but it also indicates that the hormone is the same in all types of plants.

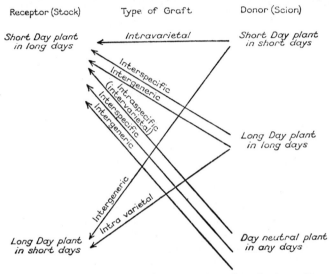

FIG. 82. Results of grafting experiments. (Based on the data of Lang)

The Effects of Auxin on Flowering. These effects are rather variable. For the most part they inhibit flowering. It is likely that its effect is on the translocation of the hormone rather than on its synthesis.

The Formation of the Flowering Hormone. One of the earliest schemes for a method of formation was put forward by Gregory and, in a somewhat modified form, this is given in fig. 83. A is the product of the high light intensity stage and it may, in excess light, be converted to a leaf-forming substance X. (There is some evidence to show that if flowering is prevented then there is greater foliar development.) In the dark period A is converted into B. The effect of interrupting the dark period depends on the extent to which the A $\rightarrow$ B reaction has taken place. An interruption at the start of the light period has little effect since there is still adequate time for resynthesis and interruption at the end of the dark period is relatively ineffective as an adequate quantity of B has been formed. References to $P_{IR} \rightarrow P_R$ refer to the phytochrome system (pages 215–16). P_R is the

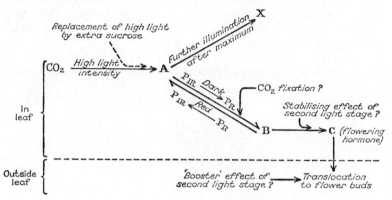

FIG. 83. Stages in formation of flowering hormone (modified after Gregory)

inactive form of the pigment and is formed from P_{IR} in the dark. The dark reaction only proceeds at a low P_{IR} level.

Long Day Plants

The grafting experiments described previously show that the flowering stimulus is the same in both long and short day plants, but it will quickly become apparent that the means by which it is synthesized must be different. Some of the characteristic features of long day plants are as follows:

Long Dark Periods are Inhibitory. This is supported by at least three lines of evidence:

(a) A plant will flower if it is kept *either* in long days *or* in short light–*short* dark cycles but it will *not* flower if kept in short light–*long* dark cycles.

(b) At lower temperatures the critical daylength is reduced and there is increased flowering. This can be explained on the basis of a temperature-sensitive *inhibition* of flowering taking place in the dark. At lower temperatures the extent of this inhibition is reduced and so the length of the critical light period is reduced or, for the same amount of light, the degree of flowering is increased. Also fitting in with this explanation is the fact that the inhibitory effect of the dark period is reduced if the low temperature is confined to the dark.

(c) In some cases defoliation results in the ability to flower under short day conditions. Since the entire plant would only flower under long day conditions it is possible that the leaves produce an inhibitory effect.

The Time of Action of the Dark Period. In theory this could be either before or after the light period (i.e. either the light could remove an inhibitor caused by the dark or the dark could remove a promoting effect caused by the light). The inhibitory effect of the dark can be stopped by supplying light of a low intensity and in most cases this low light is found to be more effective in promoting flowering if given *before* the high intensity light—i.e. the dark period inhibits the effect of the following light period.

Difference between Long Day and Short Day Plants. Other ways in which long day plants differ in their response from short day plants include the effects of light interruption of a dark period—in long day plants it *promotes* flowering—and the effect of auxin, which is to *promote* flowering.

It would be very satisfying to be able to fit these facts into the same scheme as that given for short day plants. Perhaps B is formed by an alternative light-dependent pathway which requires a higher level of P_{IR} and the inhibitory effect of darkness is to reduce the P_{IR} level.

Vernalization

The term vernalization refers to the fact that in some plants the time necessary for flowering is reduced if they are exposed to low temperatures. A great deal of work has been done with cereals where both winter varieties (sown in the autumn and flower in the spring) and spring varieties (sown in the spring and flower in the summer) are known. The winter varieties often produce a much heavier crop of flowers and therefore give a better yield than spring varieties, so that from a commercial point of view it would be advantageous to be able to plant both first and second crops with winter varieties. When this was tried, the winter varieties planted in spring did not flower in summer and we now know that this is because they did not receive a cold stimulus during germination.

The first large-scale experiments were carried out by Lysenko in 1928 although Klippart and Gassner had demonstrated the principle many years before. Lysenko's method involved heaping the grain on to the concrete floor of a barn and adding water until germination started. The barn doors were then opened to chill the grain which was periodically turned to maintain good aeration. After three months the grain was dried and stored until required for spring planting.

Investigations into the mechanism of vernalization have been carried out on several plants, in particular, winter varities of 'Petkus' rye and the biennial plant *Hyoscyamus niger* (Henbane). Henbane may well be unusual in that the effect of vernalization can be transmitted across a graft—there is certainly no evidence that this can occur in the cereals—but this is the aspect which will be considered here since it provides support for the existence of another hormone, *vernalin*.

The life-cycle of the biennial henbane normally occupies two years. During the first year a basal rosette of leaves is developed and during the second year, if vernalization has occurred, a flowering stem grows and produces a terminal flower. Vernalization is only effective after the plant has reached the rosette stage—treatment of the seed or immature embryo is ineffective—and is at least ten days old. It is thought that the stimulus is perceived by the stem tip. In the case of the *annual* henbane, no vernalization is necessary and the life-cycle is completed in one year.

Grafting experiments with henbane (summarized in table 12) suggest the existence of a hormone other than florigen.

Donor	Required for donor flowering	Treatment of donor		Recipient	Treatment of recipient		Flowering of recipient
1. *H. niger* (B)	V + LD	V	LD	*H. niger* (B)	NV	LD	Yes
2. *H. albus*	DN	LD or SD		*H. niger* (B)	NV	LD	Yes
3. *H. niger* (A)	LD		LD	*H. niger* (B)	NV	LD	Yes
4. *H. niger* (B)	V + LD		LD	*N. tabaccum*		LD	Yes
5. *Nicotiana tabaccum*	SD		SD	*H. niger* (B)	NV	LD	Yes
6. *N. tabaccum*	SD		LD	*H. niger* (B)	NV	LD	Yes

TABLE 13. Vernalization grafts

Key: A = Annual species, B = Biennial species. V = Vernalization. NV = No Vernalization. LD = Long days. SD = Short days. DN = Day neutral

In (1) an unvernalized plant in the correct day length conditions is brought into flower when a vernalized plant is grafted on to it. This could mean that the vernalized plant is transmitting a stimulus from vernalization to the unvernalized plant, but it does not rule out the possibility that florigen is being transmitted. Florigen transfer could also account for grafts (2) and (3) in which a day neutral and a long day plant, neither with vernalization requirements, induce the flowering of the unvernalized henbane. From (4) and (5) it can be deduced that the stimulus produced by the vernalized plant in long day conditions is the same as the hormone required for short day flowering plants, but in (6) a *non-induced short day plant* can transmit a stimulus which causes the unvernalized recipient on correct daylength conditions to flower. In this case there is no question of florigen being produced by the donor but the donor is able to supply the effect of vernalization, i.e. *Short day plants produce the effect of vernalization even when kept at high temperatures and in long day conditions.*

The results of these various grafts have, by some workers at least, been considered sufficient grounds for postulating the existence of the hormone, vernalin, thought to act either as a *precursor* of florigen or as a catalyst for florigen production. Some further clues as to the nature of vernalin may come from work with gibberellic acid. Heavy application of gibberellic acid to non-vernalized henbane results in the development of the flowering stem, even under short day conditions: under long day conditions these induced stems will flower although they sometimes contain a higher percentage of aborted seeds.

It is tempting to add vernalin to the list of hormones but it must be remembered that very few plants show the ability to transmit the effects of vernalization by grafting.

The Gibberellins

The first work on the gibberellins resulted from Japanese investigations into the Bakanae disease of rice caused by the soil fungus— *Gibberella fujikuroi.* One of the earliest symptoms of the infected plants is a marked elongation of the shoots and leaves. Kurosawa was able to make cell-free extracts of the fungus and show that these caused similar symptoms when injected into rice seedlings. By 1939 small samples of crystals of gibberellin A had been obtained.

After the Second World War British and American scientists took an interest in the gibberellins and one of the first results of their investigations was the development of much-improved culture techniques and the isolation of several gibberellins. So far five have been isolated, A_1, A_2, A_3, A_4 and A_5. A_3 is usually known as gibberellic acid (GA).

The empirical formulae and possible interrelationships are

$$C_{19}H_{22}O_5 \underset{-H_2O}{\overset{+H_2O}{\rightleftharpoons}} C_{19}H_{24}O_6 \underset{+2H}{\overset{-2H}{\rightleftharpoons}} C_{19}H_{22}O_6$$
$$(A_5) \qquad\qquad (A_1) \qquad\qquad (A_3)$$

$$-O \diagup\diagup +O$$

$$C_{19}H_{24}O_5 \underset{-H_2O}{\overset{+H_2O}{\rightleftharpoons}} C_{19}H_{26}O_6$$
$$(A_4) \qquad\qquad (A_2)$$

The structural formula of GA is

In terms of biological activity, GA is the most active, A_1 and A_4 are about the same and A_2 and A_5 the least active.

The most obvious effect of GA application to the plant is on shoot growth. When supplied in concentrations of 1–10 μgm per ml there is usually a very marked increase in length of the stem, mainly because of increased cell length rather than any effect on cell division. The number of internodes remains approximately constant. In the case of plants with a main stem this makes the plant longer, but if the GA is applied to a dwarf variety (e.g. dwarf garden pea) the effect is much more spectacular. All the changes described previously take place but they are associated with a change from the bushy habit to that of the normal-sized plant so that growth is mainly confined to the main axis. Apart therefore from an effect on cell elongation the GA must be producing an effect on the growing points by inhibiting the development of laterals. There is no evidence to suggest that this is a direct effect on the laterals—application of GA to a decapitated stem does not result in a suppression of laterals as occurs with auxins, although the GA could be reinforcing the action of auxin.

Apart from the effects on the stem there may also be an increased leaf area with a consequent enhancement of photosynthesis and so a nett increase in dry weight. There is no evidence for any GA effect on the root system.

Comparisons of extracts from dwarf and tall varieties of plants for GA activity do not reveal a uniform state of affairs. In maize the GA activity of extracts from tall plants was more than twice that of extracts from dwarfs, whereas in the garden pea the activities are much the same. The simplest explanation is that in peas the dwarfism is due to an inhibition of GA activity but in maize it is due to a difference in the amounts of GA.

From the genetical point of view dwarfism is often associated with a recessive mutation from the wild type (tall), sometimes only a single mutation being involved (e.g. in sweet pea). In these cases GA has little action on the tall plant but converts the dwarf into a phenotype almost indistinguishable from the tall: presumably the effect of the mutation can be expressed in terms of the loss of an ability to synthesize GA.

Many of the actions of auxins, discussed in the previous chapter, are not shown by the gibberellins. Thus GA fails to inhibit petiolar abscission or lateral bud development, and it does not promote lateral root formation or callus formation when applied to cut surfaces.

Interactions between GA and Auxins. In the previous chapter it was mentioned (page 174) that the presence of both GA *and* auxins was necessary for the normal differentiation of xylem mother cells resulting from cambial divisions. Wareing found that when the cambial activity of dormant *Acer* shoots was stimulated by the application of GA, the new cells formed on the inside of the cambium showed little lignification; when IAA was used as the stimulant a small amount of lignification was observed, but when IAA *and* GA were applied together, not only was there a greater stimulation of cambial activity but the xylem formed showed normal lignification.

This type of effect is fairly typical: in more general terms it would seem that GA is only really effective in its growth-promoting activities if IAA is present (although the converse, viz. that IAA is only effective if GA is also present, does not necessarily apply).

Since this GA effect is found with a wide range of auxins of different chemical structure (e.g. IAA, 2 : 4D), it is unlikely that it acts by any form of chemical combination to form a more stable or a more active compound. A more likely explanation is that the GA acts by reversing the effects of an auxin *inhibitor*. Evidence for this can be obtained by comparison of the growth rates of intact and excised internodes of tall and dwarf pea stems under various conditions. The results of this type of experiment are summarized below.

	Material + treatment	Growth rate	Inter-relationships
	(i) Internodes of intact plants	a	
	(ii) Excised internodes on IAA + sucrose	b	(b > a)
Dwarf plant	(iii) Excised internodes on IAA + sucrose + GA	c	(c > b)
	(iv) Internodes of intact plants + GA	d	(d ≃ b)
	(v) Excised internodes of GA-treated plant on IAA + sucrose	e	(e ≃ b)
Tall plant	(vi) Internodes of intact plant	f	(f ≃ b)
	(vii) Excised internodes + IAA + sucrose	g	(g ≃ b)

Let us assume that the small size of the dwarf plants is caused by an auxin inhibitor, present in the intact plant but not in the excised sections, which is inactivated by GA. Thus the growth rate of excised internodes, without the auxin inhibitor, shows an increase over the internodes of intact plant when supplied with IAA (i and ii). The addition of GA to the culture medium of excised stems only results in a slight further increase in growth (c > b) and could be associated with a very small amount of inhibitor in the excised internodes. When the GA is applied to intact dwarf plants, it removes the effect of the inhibitor so that its growth rate is approximately the same as the excised internodes on IAA and sucrose (iv). Similarly if internodes are excised from a GA-treated plant (v) there is no increase in growth over that of the GA-treated intact plant. If this theory is correct then there would not be any auxin inhibitor in the tall plant and consequently excision would not be expected to raise the growth rate (vii) and the intact plant's rate would be the same as that of the excised plant treated with auxin (ii).

Possible Connexions between GA and the Flowering Hormones. GA can replace the need for vernalization in cold-requiring biennials so that these plants can be brought to flower if supplied with GA while kept in long day conditions. Sometimes there are minor differences between GA effects and true vernalization, such as whether or not the flower primordia develop before or after elongation of the flower stalk, but on the whole the case for equating GA with the effect of vernalization is a strong one.

GA will promote the flowering of long day plants kept in short day conditions (i.e. it behaves like florigen) but it will inhibit the flowering of short day plants under short day conditions. The effect for long day plants is in agreement with Lang's observation that the quantity of gibberellin in a photoinduced plant of *Hyoscyamus niger* (annual) with an elongating flower stalk is markedly higher than in the non-induced plant.

In an attempt to explain this discrepancy Brian points out that in many respects the gibberellins show similar effects to those produced by either red or infra-red light through the phytochrome system (pages 215–16). These include the breaking of dormancy in light-sensitive seeds, the stimulation of expansion of leaf tissue and of course the stimulation of flowering in long day plants and its inhibition in short day plants when administered during the dark period. He suggests that one of the essential differences between the two types of plant is a sensitivity to GA level during florigen synthesis, i.e. long day plants can only synthesize florigen when the GA concentration is high and short day plants can only synthesize it when the concentration is low. Furthermore it is assumed that GA synthesis is in some way tied up with the phytochrome system.

The Status of Gibberellins as Phytohormones. There is no doubt as to the presence of gibberellins in plant tissues—they have been extracted from stems, roots, leaves, seeds and flower primordia—or that they are transported (non-polar) and produce definite effects. R. L. Jones and I. D. J. Phillips developed a diffusion technique for extracting GA from tissues on much the same principle as that used by Went for auxin extractions. As a result of the application of this method it has been possible to show that the main sites of GA synthesis are the root tip and the young leaves in the apical buds. There is also strong evidence to suggest that GA may be synthesized in seeds.

Transport of GA has been shown to take place in the xylem and in the phloem sieve tubes. The former has been deduced from the presence of GA in xylem exudate from the surface of cut stems, whereas the presence of GA in sieve tubes can be shown by extracts obtained from aphid mouthparts. It also seems likely that there is an interchange between the GA in the phloem and that in the xylem. Thus their status as phytohormones is more substantial than that of florigen or vernalin, but there have been no demonstrations of a particular gibberellin-producing region, the effects of the extirpation and the subsequent administration of exogenous GA.

Abcisic Acid (ABA)

During the last ten years three growth-regulating compounds were described—Abcisin I, Abcisin II and Dormin. The first two compounds were described by H. R. Cairns and F. T. Addicott (1964) as a result of their studies on leaf abcission. They isolated the two compounds from cotton bolls and from one of these, Abcisin II, a pure substance, abcisic acid, was prepared.

Dormin was reported by P. F. Wareing, who was studying the cause of bud dormancy in trees. He and his co-workers were able to isolate from sycamore leaves a compound which would promote the formation of resting buds in sycamore seedlings. The further purification of dormin revealed that it has the same structure as Abcisin II.

Structure of abcisic acid (ABA)

The various physiological effects of ABA can be considered under the following headings.

a. Abcisic acid and flowering

It can be shown quite clearly that the concentration of ABA in the leaves of short-day plants increases under short-day conditions. It is

well known that under similar short-day conditions long-day plants will not flower and it has been generally accepted that this may well be due to the production of a substance that is inhibitory at the apex of long-day plants. In several cases it has been possible to show that the addition of ABA, either by injection or spraying, inhibits the flowering of long-day plants under long-day conditions. In the case of Indian rye-grass it was found that the injection was most effective at about the time when the flowering stimulus (hormone?) reached the apex. The effect of flowering on short-day plants is less clear, although in some cases at least, flowering is promoted.

b. Abcisic acid and abcission

Reference has already been made to the work of Cairns and Addicott on the abcission of cotton bolls. In the process of abcission it stimulates rapid cell division, the breakdown of the middle lamella and either the production and/or activation of enzymes such as cellulases, pectinases and proteases.

c. Abcisic acid and senescence

When leaf discs are floated on a solution of ABA it is found that within a few days they lose their chlorophyll, proteins start to break down and there is a reduced synthesis of RNA. A similar effect is *not* found when the ABA is supplied to intact leaves on the tree, suggesting that in this respect there is an inhibitor of ABA activity (or, alternatively, a stimulus for RNA and protein synthesis). This may be due to the activity of a cytokinin.

d. Abcisic acid and seed dormancy

It can be shown that in some cases the length of the dormant period of seeds can be increased by the addition of ABA, and that a few seeds will not germinate at all if kept in the presence of ABA. Some seeds—such as ash—contain ABA and so presumably this acts as a germination inhibitor. As in other cases of ABA activity, the acid can easily be removed by washing in water.

e. Abcisic acid and bud dormancy

In addition to the extracts from sycamore which could be shown to produce bud dormancy, it can be shown that the addition of synthetic ABA to a variety of tree seedlings results in the inhibition of bud development and the conversion of active growing points into typical resting buds.

The interrelationship of ABA and other phytohormones

There are several examples which suggest a form of competition between different plant hormones, e.g.—

(i) ABA induced abcissions can be inhibited by the action of IAA.

(ii) There may be interactions between ABA and florigen.

(iii) ABA will inhibit the synthesis of α amylase synthesis which is stimulated by gibberellic acid:

Ethylene as a plant hormone

S. P. Burg states that 'the earliest observations on the biological effects of ethylene are hopelessly entangled in a literature describing the toxic action of illuminating gas and smoke on plant tissues'. These observations were made contemporaneously with the first work on auxins and phototropism at the end of the last century, but it was not until 1901 that Neljubow showed that ethylene was responsible for altered patterns of leaf and stem growth in peas.

A large part of the early work was concerned with observations that ethylene was involved in the ripening of fruits: thus it was shown that the gaseous products of oranges hastened the ripening of bananas and later it was demonstrated that the active principle was ethylene. Although a considerable amount of work has been done on such diverse effects as ethylene-induced abcission, tropisms, flowering, etc., in this account only its role in the ripening of fruits will be considered. The study of ethylene has made tremendous progress in the last few years with the introduction of the flame ionization gas spectrograph which provides a means of detecting very low (1 in 10^9) concentrations.

Experimental work has been done on a wide range of fruits and the following generalizations can be made—

(i) Mature unripe fruits can be ripened by the application of ethylene.

(ii) Self-ripening fruits produce ethylene during the ripening period.

(iii) Ethylene is present in the intercellular spaces before it can be detected in the external atmosphere.

(iv) The changes accelerated by ethylene consist of a rise in the respiration rate (the climacteric), followed by changes in colour (including a loss of chlorophyll), texture and sugar content.

(v) There is no ethylene synthesis in the absence of oxygen and its synthesis is slow at low oxygen concentrations.

Although a close parallel can be seen between the production of ethylene and the onset of the climacteric rise in the respiration rate, this of itself does not distinguish between cause and effect. When gas spectrographic methods are used to assay the changes in the intercellular concentration of the ethylene, simple arithmetical plots do not really help, but when log plots are made it can be seen that there is a $\times 50$ to $\times 100$ rise in the concentration of the ethylene *before* there is a detectable change in the rate of respiration.

It has already been mentioned that at low oxygen concentrations there is a retardation of ethylene synthesis, so that tomatoes and bananas can have their ripening arrested by being stored in 5% O_2 (thus avoiding the physiological damage which would result from storage under anaerobic conditions). They can be induced to ripen under these conditions by the addition of ethylene at a concentration of one part per million: accompanying this process is a production of ethylene by the fruit, so that it seems likely that ethylene synthesis is an autocatalytic process. There would also seem to be a case to be made for believing that there is a gradual increase in tissue sensitivity to the gas as the ripening process continues.

The idea of ethylene as a fruit-ripening hormone was first put forward by F. Kidd and C. West in 1933, but it is only now that we can consider the case to be substantiated. As a plant hormone it certainly has some unusual characteristics—a simple chemical structure, a gas at normal temperatures and consequently transport in the organism by simple diffusion.

11

Germination and Growth

In this chapter it is hoped to bring together many of the physiological systems that have already been described so that they can be seen to be functioning as the individual parts of an integrated whole.

Germination

The Structure of the Seed. This varies from one species to another, but in fig. 84 most of the principal features are shown although in any individual seed not all the parts will be equally developed.

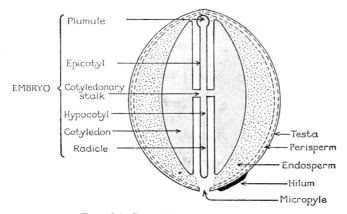

FIG. 84. Generalized structure of a seed

The seed originates from the fertilized ovule. The integuments of the ovule form the seed coat (testa) which, apart from acting as a protection for the underlying tissue, frequently restricts germination. The micropyle represents the original opening of the integuments through which the pollen tube entered and the hilum is the part where the ovule was attached by the funicle to the ovary wall. The only other direct derivative of the parental sporophyte tissue is the perisperm, formed from the nucellus. In most seeds the perisperm is poorly developed but in many members of the Caryophyllaceae it forms an important food store.

A feature peculiar to angiosperm reproduction is the phenomenon of double fertilization. One of the male pro-nuclei fuses with the egg nucleus to form the zygote and hence, by cell division, the embryo, while the other fuses with the double fusion nucleus to give the triple fusion nucleus—a triploid nucleus that forms the endosperm. This may be the main food store or it may hardly be developed. In the former case the food is absorbed into the embryo through the cotyledons and in the latter, the cotyledons usually contain the food.

The cells of the embryo appear shrunken with small vacuoles and dense cytoplasmic contents. Germination can be considered as the process by which this tissue becomes converted into an actively growing organism with a recognizable root and shoot. It is extremely difficult to say when germination finishes and growth starts; the difficulty is to some extent overcome by the definition of germination which regards it as the process of coupling respiration to growth.

According to the structural changes involved, germination can be classified into either hypogeal or epigeal. In *hypogeal* germination the epicotyl enlarges and the cotyledons and developing radicle remain below ground. In dicotyledons the plumule tip remains embedded between the cotyledons with the result that the epicotyl becomes hooked ('plumular hook'). This ensures that the soil is loosened and the delicate plumule tip is not damaged by the abrasive action of the soil. Monocotyledons on the other hand may have the first leaf developed as a protective layer around the plumule to form the coleoptile.

If the hypocotyl enlarges then the cotyledons and the embedded plumule tip emerge above ground giving the epigeal type of germination. Here the cotyledons serve to protect the plumule tip, apart from providing a food store and acting as the first photosynthesizing leaves.

Changes in Germination. The physiological changes in germination are extremely complex but it is possible to obtain an understanding of the essentials if it is remembered that

- (a) It is necessary to convert the stored food materials in the cotyledons or endosperm into structural components of new cells, particularly at the tips of the plumule and radicle.
- (b) These synthetic processes require a supply of metabolic energy which can only be supplied by aerobic respiration.
- (c) Supply of water is necessary for cell enlargement.
- (d) Water is necessary for hydrolyses and translocation.

It is evident that germination will only take place if oxygen and water are present and a suitable temperature is required so that the many enzyme-catalysed reactions can proceed at optimum, or near optimum, speeds. Some seeds also need red light for germination.

If these conditions are supplied to the seeds of cultivated plants then there is a very high probability that they will germinate, but if they are supplied to the seeds of 'wild', non-cultivated plants there is an even higher probability that they *won't* germinate. There are several possible reasons for this, but whatever the cause, there is a sound ecological reason. If all the seeds shed from a plant were ready to germinate at the same time it would be very much a case of putting 'all one's eggs in the same basket'. Thus all the seeds of a desert plant might germinate after a single shower only to be killed off in a following drought, whereas if there is some form of *regulated* germination there is a fair chance that some of the seeds will survive until more advantageous conditions obtain. The mechanism of such regulation varies—water-soluble growth inhibitors may be present in the testa and germination will only take place when they have all been leached out; many legumes have a waterproof testa which only gradually becomes permeable and in the pericarp of the tomato there is a chemical inhibitor which prevents germination of the seeds while they are still enclosed in the fruit. The seeds of parasitic plants normally only germinate if they are in the vicinity of a suitable host and it is very likely that this is because of a stimulatory exudate from the roots.

Apart from these regulatory processes there is usually a delay before the seeds can germinate. This condition of dormancy can be due to several causes—immaturity of the embryo, impermeability of the testa to water or oxygen, or the inability of seed tissues to exert a sufficient swelling force. When considering the effects of external factors on germination it is necessary to assume that the dormant period is over.

The necessity of water for germination is paramount. Water enters through the testa and the micropyle, the relative importance of the two routes varying according to the permeability of the testa. The cause of water absorption is, in the first place, imbibition caused by the presence of proteins and mucilages. This absorption is later re-inforced by an osmotic component when starches become hydrolysed to sugars. The consequences of this water absorption are fourfold:

(i) By the resultant swelling the testa is ruptured and oxygen can enter making aerobic respiration possible. In addition any CO_2 which has accumulated inside the seed (and perhaps acted as a germination inhibitor) can diffuse out, and the developing embryo can emerge.

(ii) The increased water content (in *Ricinus* from 6·5% to 92·7%) provides a reactant for the hydrolyses involved prior to trans-location of the reserve food materials.

(iii) The water provides a medium for translocation.

(iv) Vacuolation can take place, with a consequent increase in length of plumule and radicle.

The Mobilization of the Storage Material. Most seeds have fats as their main storage material although a significant number have carbohydrates. Very few have large protein stores. Fig. 85 shows the percentages of the different types of stored materials in the pea, maize and soybean.

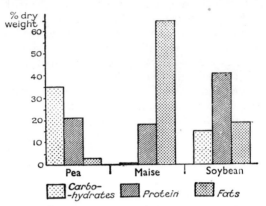

FIG. 85. Main storage compounds in pea, maize and soybean

As already indicated the purpose of these storage materials is to

(a) Provide a respiratory substrate and therefore a source of energy for endergonic reactions.

(b) Provide new materials for growth. In particular this involves the synthesis of new protein, complex lipids (e.g. phospholipids) and cellulose.

Some of the gross changes, based on Yoccum's analysis of wheat seedlings, are shown in fig. 86. In 86a there is a steady decrease in dry weight of the endosperm accompanied by increases in dry weight of the plumule and radicle. The bulk of the storage material is starch and the loss of starch from the endosperm is rapid. 86b shows the decrease in the *total* dry weight of the seedling during germination. This is not a steady change but increases sharply from the ninth day. The significance of the loss of dry weight lies in its value as an indication of the rate of respiration.

Changes in the fat and nitrogenous (mainly protein) constituents are shown in 86c and d respectively. Although the decrease in the endosperm fat content is accompanied by a rise in the fat level of both plumule and radicle, by the ninth day the amount of fat present in the plumule exceeds the amount originally present in the endosperm.

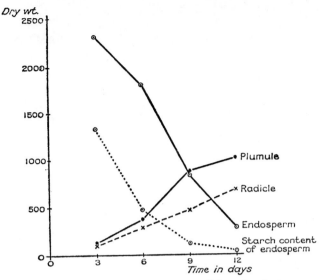

(*a*) Dry weight changes of plumule, radicle and endosperm

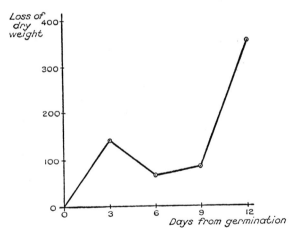

(*b*) Loss of dry weight of whole seedling

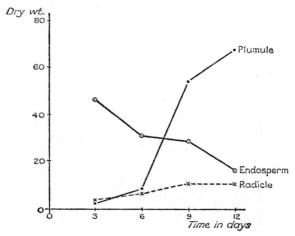

(c) Change in fat content of plumule, radicle and endosperm

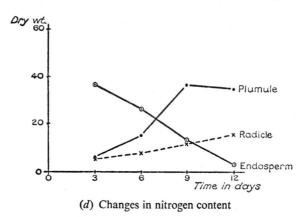

(d) Changes in nitrogen content

FIG. 86. Dry weight changes of germinating wheat seedlings (based on data of Yoccum). All results in mg. per 100 seeds or parts of seeds

This can only be explained in terms of fat formation, probably from carbohydrates (note the different scales of the ordinates of 86a, b and c). The nitrogenous changes show the same general pattern as previously noted—a decrease in the endosperm with a concomitant rise in the radicle and plumule. As in the case of fats there is a rise in the *total* value by the ninth day. This is accounted for by the absorption of nitrates through the developing radicle.

The mobilization of starch is brought about by amylase catalysed hydrolyses (for further details see chapter 1, page 15). The resulting maltose is further hydrolysed to glucose under the influence of maltase. Glucose may be transported from the endosperm as such (or more likely as glucose 1 phosphate) or it may be converted into sucrose.

Apart from the effect of glucose on osmosis it may be involved in cellulose formation or broken down via triose phosphates, diphosphoglyceric acid, pyruvic acid and the Krebs' cycle to form CO_2, H_2O and about $38 \sim \text{(P)}$ for every gram molecule utilized (aerobically). Fat synthesis can take place by the condensation of molecules of acetyl coenzyme A to form fatty acids, by the reduction of dihydroxyacetone phosphate to form glycerol and then the condensation of fatty acids and glycerol to form fats. All the available evidence points to the importance of the Embden–Meyerhoff pathway for glucose breakdown (i.e. the pathway described in chapters 1 and 8) rather than the alternative pentose phosphate shunt.

Utilization of fats will be along the lines indicated in chapter 6. The initial reactions are hydrolyses catalysed by lipases to form fatty acids and glycerol. These products will then be translocated to the growing regions and

(a) Be involved in phospholipid synthesis.
(b) Be used as an energy source.
(c) Contribute to carbohydrate syntheses (e.g. of cellulose).
(d) Form carbon skeletons for amino acid syntheses.

The relative importance of these four reactions will depend on the type of seed. Phospholipid synthesis will remain fairly constant but the fats will make only small contributions to (b), (c) and (d) in carbohydrate-storing seeds.

In the last three cases glycerol will enter the main metabolic pathways via dihydroxyacetone phosphate but the fatty acids will enter via acetyl coenzyme A and the Krebs' cycle. α ketoglutaric acid can give rise to glutamic acid if the radicle is sufficiently developed to allow adequate salt absorption and α ketoglutaric acid, oxaloacetic acid and pyruvic acid can take part in transamination with amino acids from protein hydrolysis. For carbohydrate synthesis from fats the Krebs' cycle is inadequate because of the irreversibility of the reaction pyruvic acid $\rightarrow$ acetyl coenzyme A. It is therefore necessary to utilize the glyoxylic acid cycle with the production of pyruvic acid.

The first stage in protein synthesis is the hydrolysis of the storage protein, under the influence of endo- and exopeptidases, to form amino acids. Some of these are translocated as such but others are deaminated and their amino groups transfered to either α ketoglutaric acid or oxaloacetic acid forming first glutamic or aspartic acid and then glutamine or asparagine (page 107). The amides are transported to the growing regions and the α keto acid residues oxidized. By a reverse series of reactions the amides lose their NH_2 groups to appropriate α keto acids synthesized *de novo*. The synthesis of proteins from the amino acids is under the direct influence of RNA.

Respiration. At the beginning of this chapter a definition of germination was given, viz. a coupling of respiration and growth. Respiration in the dormant seed proceeds at a very low level and, because of the low permeability of the testa, it is frequently anaerobic. CO_2 tends to accumulate and this inhibits germination.

When the permeability of the testa increases, either because of its hydration or its rupture, there is a marked rise in the rate of respiration and the R.Q. is greater than one (i.e. the CO_2 output is greater

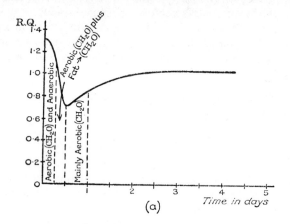

(a)

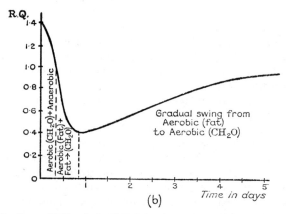

(b)

Fig. 87. Generalized graphs of R.Q.'s of germinating seeds—(*a*) with a carbohydrate store and (*b*) with a fat store

than the oxygen intake) for the first few hours as the respiration swings from anaerobic to aerobic with a carbohydrate substrate. This is followed by a fall in R.Q. to about 0·7. In carbohydrate-storing seeds the drop in R.Q. is of short duration as the small fat store is oxidized to carbohydrate, but in seeds with a fat store the drop continues to a value of about 0·4—here fat is used as a respiratory substrate *and* there is a conversion of fat to carbohydrate: it may be several days before the R.Q. eventually reaches unity (fig. 87*a* and *b*)

Temperature. The effects of temperature on a seed are rather variable. In the first place dry seeds are able to withstand large extremes of temperature and examples have been recorded in which the viability of seeds is unaffected by exposure to liquid air or to temperatures as high as 90° C. On the other hand, once the seed has imbibed water its viable temperature range becomes much more restricted. Results are usually expressed as the *minimum* temperature (below which germination will not occur), the *optimum* temperature (at which the maximum percentage germination is found) and the *maximum* temperature (above which no germination takes place). These terms must be treated with some caution because the criterion of 'no germination' varies considerably—if the percentage of germination is very low it is usually very slow and different workers adopt different standards. As an example the figures given in table 13 give values for *Zea mais*.

Minimum	Optimum	Maximum	Authority
4·8	37–44	50	Haberlandt in Stiles, *Intro. to Plant Physiol.*
8–10	32–35	40–44	Mayer and Poljakoff-Mayber, in *The Germination of Seeds*

TABLE 13

The cause of the minimum temperature is not known, but the maximum is associated with protein denaturation. The effect of temperature will manifest itself as an integration of its action on the many enzyme systems involved in metabolism and also its effect on water uptake by imbibition due both to a lowering of the viscosity of the water and an increase in the kinetic energy of the water molecules.

Two other temperature effects must be mentioned:

(i) *The effect of alternations of temperature.* In many cases the percentage germination is appreciably higher if instead of maintaining the seeds at a constant temperature they are subjected to a fluctuating temperature. For the seeds of the grass *Agrostis alba* Lehman and Aichele found

At a constant temperature of 12° C:
the percentage germination = 49
At a constant temperature of 21° C:
the percentage germination = 53
At a fluctuating temperature between 12 and 21° C:
the percentage germination = 69

The causes of this are unknown.

(ii) *Exposure to a low temperature before germination* sometimes results in an increased percentage germination, especially if the low temperature is combined with an increased humidity. The phenomenon is known as *stratification* and is associated with alterations in enzyme content and nitrogen and phosphorus metabolism in the after-ripening process. As an alternative it is sometimes possible to substitute removal of the testa for stratification but when this is done the resulting seedlings are often abnormal. A simple explanation of stratification might be that the low temperature blocks the action of a germination inhibitor with a high Q_{10}. There are some indications that the gibberellins may be involved.

Light. It has been known for many years that light can influence germination and Kinzel in 1926 divided plants into three groups on this basis:

(*a*) Seeds stimulated to germinate by exposure to light.
(*b*) Seeds stimulated to germinate by exposure to dark.
(*c*) Seeds 'indifferent to' illumination.

Most of the work on light-sensitive seeds has been done with lettuce seeds (*Lactuca sativa*) and by the use of optically pure filters it has been possible to obtain action spectra. In general light-sensitive seeds show the following characteristics:

(*a*) Germination is inhibited by blue light
(wavelength *c.* 4,500 Å)
(*b*) Germination is inhibited by infra-red light
(maximally at 7,300 Å)
(*c*) Germination is promoted by red light
(maximally at 6,800 Å)

It has also been found that the inhibition produced by infra-red (far red of some authors) can be annulled by a subsequent exposure to red, so that if an alternating series of exposures to red and infra-red is given it is the last light which is effective (table 14).

Exposure to	Germination
R	Promoted
IR	Inhibited
R–IR	Inhibited
R–IR–R	Promoted
R–IR–R–IR	Inhibited
IR–R–IR–R	Promoted

TABLE 14. Effect of red (R) and infra-red (IR) light on germination of light-sensitive seeds

This effect is due to the formation of the active form of phytochrome (P_{IR}) and is discussed in more detail on pages 215–16.

The operation of blue light inhibition is imperfectly understood. It operates at a higher intensity than the red–infra-red system and has an action spectrum which shows a second peak in the infra-red. No pigment system has yet been isolated for it.

Plant Growth

In this section the growth of the plant after the seedling stage is discussed, when it is entirely dependent on external supplies of food.

The main components of this growth are the processes of cell division, cell elongation and then cell differentiation. Superimposed on these is an overall control which determines the form of the whole plant body and the relationship of the various parts to each other.

Cell division involves mitosis and cytokinesis—viz. a quantitative division of the nucleus resulting in two nuclei each having exactly the same genetic constitution as the parent nucleus followed by the division of the cell contents into two. Cell division occurs mainly at the root and stem apices, in the formation of secondary vascular tissues (the meristematic cells being the cambium) and the formation of secondary cortical tissues by the action of the phellogen.

Apart from those of the cambium, meristematic cells are characterized by a conspicuous nucleus embedded in the centre of dense cytoplasm; a conspicuous vacuole is absent but many very small vacuoles may occur. During growth there is little increase in the total quantity of cytoplasm but a conspicuous central vacuole develops by the coalescence of the small ones. The amount of cell sap increases markedly so that at this stage there is a large uptake of water and the nucleus comes to occupy a peripheral position. The cell wall is extended in area, initially by stretching, during which very little new material is added, followed by a stage in which there is an active synthesis of new material.

Electron microscope studies by Frey Wyssling show that whereas the main part of the wall consists of dense parallel strands, the ends of the wall show an open weave. This suggests that new wall material is added at the ends, but in fact all the experimental evidence favours the view that growth takes place evenly over the whole surface of the wall. As examples of such evidence we may mention

(a) 'Feeding' experiments with $^{14}C_6H_{12}O_6$ show a uniform distribution of labelled cellulose in the walls.

(b) The use of either copper oxide dust or the position of pits as 'markers' indicate again that growth is a uniform process (i.e. the increase in distance between two adjacent pits or between two adjacent specks of copper oxide dust is the same for any position on the wall).

The course of cell differentiation will vary according to the type of tissue which is being formed. In the formation of fibres there is marked elongation of the ends of the cell in a vertical direction so that 'sliding' growth can take place between adjacent cells, and at the same time the cytoplasm forms large quantities of lignin in and under the secondary cellulose wall so that eventually the cell contents become completely isolated and death of the cell results. On the other hand sieve tube formation first involves a longitudinal division so that the nucleated companion cell is cut off from the enucleate sieve tube element whose end walls become modified, by the aggregation of plasmodesmata, to form sieve plates.

Rate of Growth. The rate of plant growth can be conveniently assessed by measuring changes in dry weight. The data can be represented graphically but care should be taken to distinguish between *growth curves* in which the dry weight of the plant (W) is plotted against the time (T) and *growth rate curves* in which the rate of growth of the plant (dW/dT) is plotted against the time (T). These are shown in figs. 88a and b.

In the growth curve there is an initial loss in dry weight (I) due to the consumption of the carbohydrate or fat reserves in respiration and, to a lesser extent, condensations occurring in synthetic reactions. In the second phase (II) there is a rapid increase in dry weight, mainly accounted for by photosynthesis (80–90%) but also by the absorption of mineral salts.

After this active growth stage (sometimes referred to as the 'Grand Period of Growth') there is a loss of dry weight because the rate of respiration is greater than the rate of photosynthesis (III—senescence). Finally only the dead skeleton of the plant remains and there are no longer any changes in weight.

When the rate of growth is considered (i.e. dW/dT for the grand

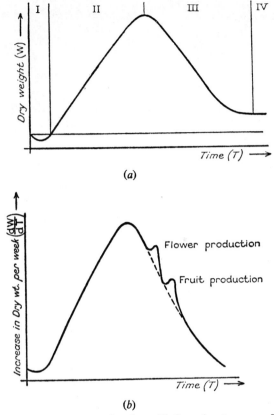

FIG. 88. (a) Growth curve of a plant. (b) Growth rate curve of maize.
(After Briggs, Kidd and West, 1920)

period of growth) a sigmoid curve is obtained which is followed by a fall in growth rate. In Briggs' generalized curve for maize, two subsidiary increases in growth rate occur, the first associated with flower production and the second with fruit formation.

Factors Affecting the Growth of Plants. These can be divided into internal and external factors. The former includes the necessity for meristematic tissues, the correct balance of hormones and the appropriate supply of vitamins (page 217). Only the barest details are given of the external factors.

A supply of water for

(a) Inhibition and rupture of the testa in germination.
(b) Vacuolation and cell elongation.
(c) Turgidity and maintenance of a rigid structure.

(*d*) A medium for translocation.
(*e*) Photolysis.
(*f*) Hydrolyses.

A *supply of oxygen* for aerobic respiration and the supply of energy for endergonic reactions.

A *supply of macro- and micronutrients* for

(*a*) Participation in molecular structure (e.g. chlorophyll, cytochrome, nucleic acids, membrane systems).
(*b*) Ionic balances.
(*c*) Electron transfer systems.
(*d*) Functioning of enzyme reactions.

Carbon dioxide, in non-toxic concentrations, for photosynthesis and flowering.

A *suitable temperature,* or variations of temperature, for

(*a*) Stratification phenomena in seed germination.
(*b*) Optimum activity of enzyme-catalysed reactions.
(*c*) Vernalization.

Absence of toxic compounds (as found, for example, in peat soils)

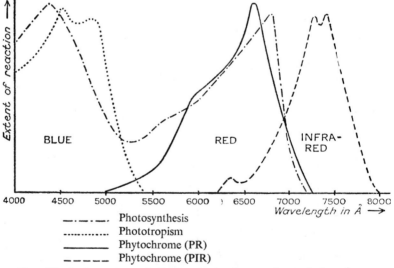

_ . _ . _ .   Photosynthesis
. . . .   Phototropism
_____   Phytochrome (PR)
_ _ _ _ _   Phytochrome (PIR)

F IG . 89. Action spectra of photosynthesis, phototropism and phytochrome effects

or of growth inhibitors. A good example of a toxic substance is aluminium. This is toxic to calcicoles when grown on acid soils.

Adequate radiation involving the following spectral components (see fig. 89).

(a) *c.* 4,350 Å for the synthesis of chlorophyll from protochlorophyll.

(b) *c.* 4,400 Å and 6,800 Å for maximum photosynthesis.

(c) *c.* 4,500 Å to 4,900 Å for phototropic curvatures and so appropriate orientation of plant organs with respect to light.

(d) *c.* 6,600 Å for red light effects associated with the phytochrome system.

(e) *c.* 7,300 Å for infra-red effects associated with the phytochrome system.

Phytochrome effects have already been discussed in several connexions. Some of the main ones are given in table 15. When a reaction is described as promoted (or inhibited) by red light then it is also inhibited (or promoted) by infra-red light.

Promoted by red *(inhibited by infra-red)*	*Promoted by infra-red* *(inhibited by red)*
Flowering of long day plants Leaf development Germination Pigment development (e.g. anthocyanin)	Flowering of short day plants Stem elongation

TABLE 16

The close similarity of the action spectra of these different processes combined with the fact that if alternating exposures to red and infra-red* are given it is the last illumination which is effective (see effects of light on the germination of lettuce seeds, page 210), indicates that the same pigment is involved, capable of existing in two forms, P_R, absorbing light maximally at 6,600 Å whereupon it is transformed into P_{IR}, capable of absorbing light maximally at 7,350 Å and being reconverted into P_R.

$$P_R \underset{\substack{\text{Infra-red (max. at 7,350 Å)} \\ \text{(or slowly in the dark)}}}{\overset{\text{Red (max. at 6,600 Å)}}{\rightleftarrows}} P_{IR}$$

It is generally accepted that P_R is physiologically inactive and represents the storage form, while P_{IR} is physiologically active.

A partially purified phytochrome has been isolated from dark

* In many accounts, particularly of American authors, the term *far red* is used instead of infra-red. Arguments have been advanced to justify each term, but in this text infra-red is used as it is likely to be more familiar to the reader.

grown five-day-old corn shoots and this shows the reversible absorption postulated above. There is no real idea as to the mode of action of P_{IR}—perhaps it is an enzyme participating in a master reaction—and the situation is complicated by the fact that whereas the phytochrome effects can be caused by light of low intensity there is a certain amount of evidence to suggest that high light intensities may also be involved in some morphogenic reactions. This high light reaction is thought to have twin peaks in its action spectrum, one in the blue and the other in the infra-red.

The Formation of Organs—Tissue Culture Experiments. The idea of tissue cultures was first advanced by Haberlandt at the beginning of this century but it was not until 1922 that Kotté and Robbins independently developed the technique of growing isolated tissues which were either in themselves meristematic or were capable of becoming so under the conditions prevailing in the culture medium. The principle is to grow isolated tissues under strictly aseptic conditions in the dark either on a solid, or in a liquid, culture medium containing all the necessary chemicals. An important aspect of this work is to develop *minimal* culture media and in this way pinpoint the essential nutrients (note that it is not the same as supplying only the essential micro- and macronutrients). The length of time for which growth can be maintained in an individual culture is limited but if frequent transfers ('passages') are made, under sterile conditions, to fresh culture media, then in some cases unlimited growth can be obtained.

In addition to a supply of essential amino acids, mineral salts and sugars it was found that a supply of chemicals found in yeast extracts was necessary for root growth. Separation of the components of the extracts showed that the active compounds were members of the vitamin B complex—thiamine, pyridoxine and nicotinic acid. Since it is not necessary to supply the *intact* plant with these, it is a logical conclusion that under natural conditions they are synthesized in the shoot and translocated to the roots. This has already been mentioned on page 157 where it was pointed out that the vitamins show all the characteristics of phytohormones. (See also table 16.)

Skoog's work on tobacco callus cultures resulted in the isolation of *Kinetin* by Miller in 1955. In these cultures it was found that callus was formed at the morphologically lower end of the tissue when supplied with sucrose, glycine, vitamins and mineral salts in an agar medium. The growth soon slowed and stopped but could be started for a further short period by the addition of IAA. It was concluded that there was a stimulus present in the original tissue (stem segments) which had quickly been used. The addition of either coconut milk or yeast extract with the IAA resulted in new and continued cell division.

Vitamin	Chemical name	Function
Pro-vitamin A	β Carotene	Light absorption (photosynthesis). 'Screening' in phototropism?
B_1	Thiamin	When phosphorylated, co-carboxylase
B_2	Riboflavin	Part of structure of FADN. Photoinactivation of IAA?
	Nicotinic acid	As an amide, part of structure of DPN and TPN
B_6	Pyridoxine	As a phosphate, part of structure of Co.A
	Pantothenic acid	Part of structure of Co.A
H	Biotin	Part of coenzyme for decarboxylation of oxaloacetic acid and for deamination of aspartic acid
C	Ascorbic acid	Terminal oxidation (with ascorbic acid oxidase)

TABLE 16. Summary of the role of some vitamins

Preliminary analyses showed that the effect was most likely to be due to a purine but trials with all the common purines were unsuccessful. However *old* DNA was effective and this was used as a starting point of a series of extractions which finally isolated the actual compound, kinetin (6 furfurylaminopurine).

Kinetin

Among the biological effects of this compound are, in the presence of IAA, the acceleration of mitoses in tobacco callus cultures and in pith cultures. In stem sections of pea (*Pisum sativum*) it overcame the inhibiting action of the apical bud on the development of lateral buds and also the inhibitory action of IAA applied to the apex indicating that these processes are the result of a balance of kinetin and IAA.

Of more particular interest to the problem of organ formation is the effect of varying the ratio (concentration of kinetin) : (concentration of IAA) in the culture medium. As will be seen from table 17, a low ratio promotes root initiation, a higher ratio the growth of undifferentiated callus tissue and an even higher ratio the formation of leafy shoots.

| Amount (mgm) per litre | | Development of callus tissue |
IAA	Kinetin	
2·0	0·02	Abundant root formation
2·0	0·10	High growth rate of undifferentiated callus
2·0	0·50	Formation of leafy shoots. Inhibited root development

TABLE 17

In conclusion it must be emphasized that the growth of the entire plant is the result of a complex interaction of internal and external factors. The internal factors are, in the first place, under the control of the plant's own genotype almost certainly operating as a result of a regulated production of enzymes. The extent to which these can operate is governed by the availability of suitable chemicals and it is here that the external factors play their important role. As a consequence of enzyme actions the internal environment will be in a state of ordered flux and this will, through the transport system, cause far-reaching effects in other parts of the plant with concomitant changes in their external environment.

Modern physiology, working in partnership with cytogenetics and chemistry, has gone a long way towards unravelling these processes, but there is still a vast amount of work to be done with ample opportunities for the student with a sound working knowledge of the basic scientific disciplines and the ability to apply this knowledge to new problems.

Bibliography

General

Introduction to the Principles of Plant Physiology (2nd edn)	W. Stiles	Methuen	1951
Plant Physiology (4th edn)	M. Thomas, S. L. Ransom, and J. A. Richardson	Churchill	1955
The Growth of Plants	G. E. Fogg	Pelican	1963
Plant Physiology	ed. by F. C. Steward	Academic Press, New York	1959 onwards
The Experimental Basis of Modern Biology	J. A. Ramsay	Cambridge	1965

Water Relationships

Water in the Physiology of Plants	A. S. Crafts, H. B. Currier and C. R. Stocking	Chronica Botanica	1949
Plant and Soil Water Relationships	P. J. Kramer	McGraw-Hill	1949
Plant Physiology, Vol. II	ed. by F. C. Steward	Academic Press, New York	1959
The Water Relationship of Plants	ed. by A. J. Rutter and F. H. Whitehead	Blackwell	1963
Movement of Water in Plants	G. E. Briggs	Blackwell	1967
Water and Plant Life	W. M. M. Barron	Heinemann	1967

Metabolism

Dynamic Aspects of Biochemistry (4th edn)	E. Baldwin	Cambridge	1964
A Guide Book to Biochemistry	K. Harrison	Cambridge	1959
Plant Metabolism (2nd edn)	G. A. Strafford	Heinemann	1963
Intermediary Metabolism in Plants	D. D. Davies	Cambridge	1961
Nitrogen Metabolism in Plants	H. S. McKee	Oxford	1962
Photosynthesis	R. Hill and C. P. Whittingham	Methuen	1953
Photosynthesis	E. I. Rabinowitch and Govindjee	John Wiley & Sons Inc.	1969
Physiological Aspects of Photosynthesis	O. V. S. Heath	Heinemann	1969
Plant Respiration	W. O. James	Oxford	1953

Respiration in Plants (2nd edn)	W. Stiles and W. Leach	Methuen	1952
Plant Biochemistry	D. D. Davies, J. Giovonelli, T. Ap Rees	Blackwell	1964

Salt Absorption

Electrolytes and Plant Cells	G. E. Briggs, A. B. Hope and R. N. Robertson	Blackwell	1961
Mineral Salt Absorption in Plants	J. F. Sutcliffe	Pergamon	1962
The Absorption of Solutes by Plant Cells	D. H. Jennings	Oliver and Boyd	1963

Growth and Germination

Plant Growth Substances (2nd edn)	L. J. Audus	Leonard Hill	1959
Light and Plant Growth	R. van der Veen and G. Meyer	Phillips Technical Library, Eindhoven	1959
The Germination of Seeds	A. M. Mayer and A. Poljakoff-Mayber	Pergamon	1963
Apical meristems	F. A. L. Clowes	Blackwell	1961
Plant Growth and Development	A. C. Leopold	McGraw Hill	1964

Reviews

Biological Reviews of the Cambridge Philosophical Society
Scientific American   (published by W. H. Freeman)
Endeavour   (published by I.C.I.)
New Scientist
Annual Reviews of Plant Physiology   (published by Annual Reviews Inc.)

Index

ABCISIC ACID (ABA), 197–9
Abscission, 176, 197
Absorption spectrum, 86, 87, 165
Acetaldehyde, 17, 136, 139
Action spectrum, 87, 89, 165, 214, 215
Activated diffusion, 77
Activators, 12
Active uptake, 47
Adding enzymes—see Lyases
Aerobic respiration, 27, 34, 133–52
Alanine, 7, 18, 104, 117
Alcohols, 19, 21, 136, 139
Aldolase, 17
Aldoses, 1
Amides, 105, 107
Amino acids, 7, 15, 102, 114
Aminopeptidase, 16
Ammonia, 17, 102, 105, 112
Amylase, 15, 199, 206
Amylopectin, 4
Amylose, 4
Anacystis, 89
Anaerobic respiration, 20, 137
Anion respiration, 78
Anti-auxin, 173
Apo-enzyme, 12
Arginine, 7, 108, 117
Ascorbic acid, 19, 142
Asparagine, 105
Aspartic acid, 7, 104, 117
ATP, 24, 27, 77, 119, 127, 134, 182
Autolysis, 112
Auxin, 47, 153–83, 191, 194, 218
Avena, 155

BACTERIA, 110–13
Bacteroids, 112
Boron, 78, 130

CALCIUM, 81, 127
Cambium, 174, 195
Carbohydrase, 15
Carbohydrates, 1–6, 117–22, 146, 204
Carbon dioxide, 35, 83, 92, 97, 136
  145, 186
Carboxylase, 17
Carboxypeptidase, 16
Carotene, 86, 165
Carrier, 79
Catalase, 19

Cellulose, 6, 95, 181
Chlorella, 88, 89, 92
Chlorophenoxyacetic acid, 159
Chlorophyll, 85–92, 126
Chlorophyll a$_6$ 70, 90
Chlorosis, 126
Citrulline, 108
Clinostat, 169
Cobalt chloride, 55
Coding, 115–17
Coenzyme A, 29, 120
Coenzymes, 21, 28, 103, 135, 142
Cohesion theory, 64–8
Celeoptile, 153
Common path theory, 141
Companion cell, 77
Compensation point, 35
Copper, 131
Cryoscopic, 47
Cyanide, 141
Cyanophyceae, 90, 110
Cystine, 129
Cytochrome, 21, 131, 141
Cytochrome f, 90
Cytochrome b$_6$, 90
Cytokinesis, 174, 211

DARK REACTION, 83, 92–5
Daylength, 184
Day neutral, 185
Deficiency disease, 125
Dehydrogenase, 18
Denitrification, 113
Desoxyribose nucleic acid, 3, 115–17
Diffusion, 31–5
Diffusion gradient, 33
Diffusion pressure deficit, 43–50, 60,
  179
Dinitrophenol, 182
Dioses, 1
Dipeptidase, 16
Dipeptide, 8
Disaccharide, 3
Donnan Equilibrium, 52
Dormin, 197 (see also Abcisic Acid)
Dwarf varieties, 194

EDGE EFFECT, 32
Elasticity, 179
Electro-osmosis, 48

Emerson, R., 87
Endergonic, 24
Endodermis, 50
Endopeptidase, 16
Energy, 24–7, 83, 143, 207
Enhancement effect, 88
Enzymes (general), 9–14
Erepsin, 16
Ethylene, 199–200
Etiolation, 107
Exergonic, 24
Exopeptidase, 16

FADN, 21, 23
Far red, 88 (*see also* Infra-red)
Fats, 6, 14, 117–22, 145, 204
Fehling's test, 2
Fermentation, 134–7
Ferredoxin reducing substance, 90
Florigen, 184–91
Free Space, 53
Fructosans, 4
Fructose, 2
Fruit growth, 175
Fumarase, 16
Fumaric acid, 16
Fungi, 81, 112, 134

GALACTOSE, 1, 2
Geotropism, 168–74
Germination, 201–11
Gibberellic acid, 193
Gibberellins, 157, 193–7
Glucanases, 181
Glucose, 1, 15, 28, 134, 207
Glutamic acid, 7, 18, 104–7, 111
Glutamine, 105–7
Glyceraldehyde, 2
Glycine, 7, 117
Glycollic acid, 41, 152
Glyoxylic acid cycle, 121, 152
Growth curves, 213
Guanosine triphosphate, 6
Guard cells, 39

HEPTOSES, 1
Heteroauxin, 158
Hexokinase, 17
Hexose phosphate, 74, 127, 134
High energy phosphate, 25
Hill and Bendall, 90
Holoenzyme, 12
Hormones, 153–200
Humidity, 56
Hydrogen acceptor, 21, 128, 142

Hydrolases, 14
Hydroxylamine, 102

IMINO ACID, 104
Incipient drying, 58
Indolyl acetic acid, 158 (*see also* Auxin)
Infra-red, 210, 215 (*see also* Far red)
Insectivorous plants, 102, 107
Interfacial theories, 77
Inulin, 4
Iron, 23, 85, 128
Isoelectric point, 9
Isomerizing enzymes, 18

KETOGLUTARIC ACID, 18, 104, 111
Ketoses, 1
Kinetin, 216
Krebs' Cycle, 26, 30, 95, 121, 144

LACTIC ACID, 21, 136
Lateral buds, 177, 194, 198
Lateral transport, 74, 163
Leghaemoglobin, 112
Lenticels, 48, 54
Light, 35, 40, 57, 85–91, 99, 151, 153, 184, 210, 215
Light reaction, 85–91
Lipase, 14, 118, 207
Long day plants, 190–1
Lyases, 16

MACRONUTRIENTS, 123–30
Magnesium, 86, 128
Malic acid, 16, 30, 121, 145
Malonic acid, 13
Maltase, 15, 207
Maltose, 3, 15, 206
Manometric methods, 37, 145
Mass flow, 74–8
Mesophyll, 34, 65, 74
Michaelis, 13
Micronutrients, 130–2
Mineral nutrition, 123–32
Mitosis, 132, 174, 177, 211, 217
Molybdenum, 132
Monasaccharides, 1
Mycorrhiza, 81

NAPHTHYLACETIC ACIDS, 159
Navicula, 89
Nitrate, 102, 113
Nitrobacter, 112
Nitrogenase, 112
Nitrogen cycle, 112
Nitrosomonas, 112

OILS, 6 (*see also* Fats)
Oligosaccharides, 3
Optical isomerism, 2, 12
Optimum pH, 10, 118
Optimum temperature, 10, 100, 209
Organ formation, 217
Ornithine cycle, 108
Osmosis, 40, 75, 179, 203
Osmotic pressure, 41, 179
Oxidases, 18, 131, 142
Oxidation, 19, 142
Oxidation–Reduction potentials, 90
Oxidative phosphorylation, 25
Oxido-reductases, 18
Oxygen, 20, 35, 86, 101, 142, 203

P ENZYME, 4
Parthenocarpy, 175
Passive absorption, 68
Pentose shunt, 207
Pentoses, 1, 92
Pepsin, 16
Peptide linkage, 8
Peroxidase, 13, 19
Peroxisomes, 152
Phloem, 71, 161, 196
Phosphoglyceraldehyde, 1, 17, 29, 95, 136
Phospholipids, 118
Phosphorylation, 27
Photoinactivation, 165–7
Photoinduction, 185
Photophosphorylation, 27, 85
Photorespiration, 151
Photosynthesis, 83–101
Photosystems I and II, 90
Phototropism, 153–68
Phycocyanin, 89
Phycoerethrin, 89
Phytochrome, 211, 215
Phytohormones—*see* Hormones
Plasmalemma, 42
Plasmolysis, 44
Plasticity, 179
Plastocyanin, 90
Plastoquinone, 90
Pollen, 132, 175
Polypeptides, 8, 15, 114
Polysaccharides, 3, 95
Porometer, 59
Porphyridium, 89
Potassium, 126
Potometer, 55
Prosthetic group, 12
Protein, 7, 15, 114, 146, 204

Protoplasmic streaming, 183
P 700, 90
Pyruvic acid, 18, 29, 104, 136

Q ENZYME, 4
Quantum yield, 88

R ENZYME, 15
Red light, 86, 186, 211
Reductase, 103
Relative transpiration, 54
Respiration, 133–52
Respiratory quotient, 145, 208
Riboflavin, 166
Ribonuclease, 115
Ribonucleic acid, 114–15
Ribosomes, 114
Ribulose diphosphate, 92
Ringing, 63, 70
Root hairs, 49, 110
Root nodules, 110
Root pressure, 48

SALT ABSORPTION, 78–82
Salts, 50, 100, 123, 150
Sedoheptulose, 94
Seed structure, 201
Semipermeable membrane, 42
Short day plants, 185–90
Sieve plates, 76, 212
Soil water, 48, 52, 57
Specificity, 11, 114
Starch, 4, 5, 40, 74, 95, 204
Stomata, 33, 38, 54, 58–62
Storage material, 74, 106, 204
Stratification, 210
Succinic acid, 13, 22, 121
Succinic dehydrogenase, 22
Sucrose, 5, 74, 95
Suction pressure—*see* Diffusion pressure deficit
Sulphur, 129
Sulphur bacteria, 85
Surface area/volume, 34

TEMPERATURE COEFFICIENT, 9, 84
Temperature and enzymes, 10
Temperature and germination, 210
Temperature and photosynthesis, 100
Temperature and respiration, 148
Terminal oxidase, 23, 131, 142
Tetrasaccharide, 3
Tetrose, 1, 94
Tissue culture, 216
Tonoplast, 42

Transamination, 17, 104
Transferring enzymes, 17
Translocation, 70–82
Transpiration, 54–63
Transpiration stream, 63–9
Trioses, 1
Trisaccharides, 3
Trypsin, 16
Tryptophan, 7, 117, 160
Turgor, 43, 128

UREA, 108
Uridine triphosphate, 5

VACUOLATION, 178, 204, 211
Ventilation, 34

Vernalin, 193
Vernalization, 191–3
Vitamins, 158, 217
Volume—*see* Surface area

WALL PRESSURE, 43, 179
Water absorption, 48, 203
Water culture, 123
Weed killers, 177
Wind velocity, 56

XANTHOPHYLL, 86
Xylem, 63, 71, 79, 195

ZYMASE, 137